MILANESE ENCOUNTERS

Public Space and Vision in Contemporary Urban Italy

In Milan, a city driven by fashion and design, visibility and invisibility are powerful forces. *Milanese Encounters* examines how the acts of looking, recognizing, and being seen reflect social relations and power structures in contemporary Milan.

Cristina Moretti's ethnographic study reveals how the meanings of Milan's public spaces shift as the city's various inhabitants use, appropriate, and travel through them. Her extensive fieldwork covers international migrants, social justice organizations, and middle-class citizens' groups in locations such as community centres, abandoned industrial areas, and central plazas and streets. Contributing to the literature on urban anthropology and European spatial politics, *Milanese Encounters* demonstrates ethnography's potential to shed light on the complex connections and divisions within modern cities.

CRISTINA MORETTI teaches in the Department of Sociology and Anthropology at Simon Fraser University.

ANTHROPOLOGICAL HORIZONS

Editor: Michael Lambek, University of Toronto

This series, begun in 1991, focuses on theoretically informed ethnographic works addressing issues of mind and body, knowledge and power, equality and inequality, the individual and the collective. Interdisciplinary in its perspective, the series makes a unique contribution in several other academic disciplines: women's studies, history, philosophy, psychology, political science, and sociology.

For a list of the books published in this series see p. 295.

CRISTINA MORETTI

Milanese Encounters

Public Space and Vision in Contemporary Urban Italy

UNIVERSITY OF TORONTO PRESS
Toronto Buffalo London

© University of Toronto Press 2015
Toronto Buffalo London
www.utppublishing.com

ISBN 978-1-4426-4964-4 (cloth)
ISBN 978-1-4426-2699-7 (paper)

Library and Archives Canada Cataloguing in Publication

Moretti, Cristina, 1970–, author
Milanese encounters : public space and vision in contemporary urban
Italy / Cristina Moretti.

(Anthropological horizons)
Includes bibliographical references and index.
ISBN 978-1-4426-2699-7 (pbk.).   ISBN 978-1-4426-4964-4 (bound)

1. Public spaces – Italy – Milan.   2. Social groups – Italy –
Milan. 3. Visualization – Social aspects – Italy – Milan.   I. Title.
II. Series: Anthropological horizons

HT185.M67 2015   307.760945'211   C2014-907265-1

University of Toronto Press acknowledges the financial assistance to its
publishing program of the Canada Council for the Arts and the Ontario
Arts Council, an agency of the Government of Ontario.

Canada Council   Conseil des Arts
for the Arts   du Canada

ONTARIO ARTS COUNCIL
CONSEIL DES ARTS DE L'ONTARIO
an Ontario government agency
un organisme du gouvernement de l'Ontario

University of Toronto Press acknowledges the financial support of the
Government of Canada through the Canada Book Fund for its
publishing activities.

Contents

Figures

Table

Acknowledgments

This book is the result of a very long process, and many people along the way offered insight, help, and encouragement. It is impossible to mention them all here or to describe fully the many ways in which they supported this project.

Thank you to all of the people and organizations in Milan who shared their insights about the city with me: Francesca, Maria Anacleta, Don Felice, the Schuster Youth, Mohamed Ba, Alice, Giacomo, Eliza, Don Pietro, Riva, Marta, Carlo, Ottavia, Alberto, Pietro, Carlo Giorgi, the vendors and the writers of *Terre di Mezzo*, Paola Rottola, Caro, Carlo B., Carla, Mario, Licia, Luciana, and many others. Thank you to all the elders who shared their memories of Milan with me, to the young women who commented on the *struscio* (regional term for the promenade), and to the people of the Leoncavallo, Casa Loca, Naga, the Zenobia Workshop, and the Stecca degli Artigiani. Thank you for your generosity, your comments, and your teachings. I am also deeply indebted to the Department of Anthropology at the University of Milan-Bicocca, and especially to Ugo Fabietti and Roberto Malighetti, for their knowledge and hospitality.

Many people were essential in my writing this ethnography, and I am grateful to everyone who aided me in this process. A special thank you to Dara Culhane, without whom this research and this book would not have been possible and who is still my mentor in many ways; to Adrienne Burk, who has taught me a lot about public space and who has been a supporter of this project all along; and to Nicholas Blomley, Stacy Leigh Pigg, Emanuela Guano, and Kirsten Emiko McAllister for their caring guidance and critical commentaries. Thank you!

Many people at the University of Toronto Press helped me to transform the original manuscript into a book. I want to thank especially Douglas Hildebrand for his wonderful guidance, and three anonymous reviewers of the University of Toronto Press, whose advice, critiques, and suggestions greatly contributed to the manuscript. I am also grateful to Anne Brackenbury and Rae Bridgman for reading an earlier version of this book, and to Kate Baltais for editing the manuscript.

In addition to the people mentioned above, many other persons also offered insightful contributions to the research and / or comments on my work along the way, and for this I am very grateful. They include Daniel Makagon and the reviewers of *Liminalities*, Andrew Lyons and the reviewers of *Anthropologica*, Judith DeSena and the reviewers of *Gender in an Urban World*, Daniele Cologna, Maurizio Baruffi, Rima Noureddine, Sara Ferrari, Anna Morelli, Francesca Sala, Monica Muraro, Leslie Robertson, and the participants of the CASCA 2009 Ethnography Out of the Box panel. All omissions and errors are mine only. I thank Julie Cruikshank for always believing that this work could eventually become a book.

I also wish to thank my co-founders of the Centre for Imaginative Ethnography: Dara Culhane, Adrienne Burk, Denielle Elliott, and Magda Kazubowski-Houston for many inspiring discussions on anthropology, and for their friendship.

Several people helped me with the visual aspect of the book. Thank you to Enrica Sacconi for photographing the Duomo Piazza at night, the sign in the Galleria Vittorio Emanuele, and the placards in Piazza Fontana, and for drawing the maps in chapter 8; to Aurelio Bonadonna for his photographs of the Duomo, the Sforza Castle, Via Mercanti, the Armani advertisement, and two store windows; to Monica Muraro for her photograph of the Security Package demonstrations; to Daniel Say and Junas Adhikary for their help with the map in the Introduction; and to Junas Adhikary for helping me with the manuscript overall.

A special thank you to all of my family (including my parents/parents-in-law, my husband, my children, and my extended family) who helped me in this very long process in more ways than I can describe here. They provided moral and material support – from encouraging me, sharing stories, and nourishing my body and soul, to providing child care and a place to stay, helping me print the manuscript, complete bibliographies, and proofread the text. Thank you also to my children Akash and Anna for following me to Milan during my research and for "being so awesome" (as Akash likes to say).

I am grateful to the Social Sciences and Humanities Research Council of Canada for the doctoral grant that supported the initial period of this research, and to Emerald for granting me permission to reprint the material from "A Walk with Two Women: Gender, Vision and Belonging in Milan, Italy," in Judith N. DeSena (ed.), *Gender in an Urban World* (Research in Urban Sociology, vol. 9), Emerald Group Publishing Ltd, 2008, pp 53–75.

Last but not least, a big thank you also to all my friends in Italy and in Canada for their encouragement and their friendship.

MILANESE ENCOUNTERS

Public Space and Vision in Contemporary Urban Italy

Introduction

A few months after the 2011 revolutions in Egypt and Tunisia, another, albeit much smaller uprising, was taking shape in Italy. In several cities, spirits were high because of the hope that municipal elections would change the political destinies of the country and oust the increasingly unpopular prime minister, Silvio Berlusconi. (Indeed, although not known at the time, this was the beginning of a political landslide that caused him to resign six months later.) In Milan, the largest city of northern Italy and a Berlusconi stronghold, these hopes engulfed streets and open spaces, reminding everyone about the power of the piazza.

As I take a walk on a quiet evening during this period – it is May 2011 – an insistent sound grips my attention. I am close to the Central Station, the area of the city I grew up in, and where I often return. I hear music in the distance, mixed in with voices, coming from the direction of the train station: but how could it be? The area just next to the massive white building is usually sombre and silent, a busy corridor where traffic and people rush past without attempting conversation, or a smile. Following the sound, I reach the main space in front of the station and stare in disbelief. A thousand people are gathered there in celebration. There is a concert performance on a stage, and between songs speakers are encouraging everyone to vote for the centre-left coalition led by Giuliano Pisapia. I am certainly not surprised that people are using an open urban space to advocate for a political candidate during an election campaign. What strikes me, however, is that they have chosen this particular area, which almost never becomes a place for gatherings, rallies, or events. Yet, how easily this place has become a piazza in its own right! Some people have come for the politics, some for the free music, and some for a new space to party. Irrespective of the reasons, it has become an intense site of interaction, where strangers speak to each other, vendors try to sell their wares, families stroll with their children, and groups of friends sit in clusters on the grassy areas.

I think back to another incident that similarly created a new layer of meaning over familiar spaces, this time in the winter of 2008. I remember looking with surprise at the images, then flooding much of the Italian media, of hundreds of people in piazzas and courtyards, all over Italy, listening attentively to lectures on physics, history, and political science. In one such photograph, young people were sitting close together on stone steps under the open sky, surrounded by the brick walls of old buildings as if within a warm embrace; in front of them, an older man was speaking into a microphone from a desk that had been placed right in the middle of the piazza.

Like the people in these images, thousands of students and teachers in all cities in Italy had walked away from their classrooms to criticize a cut to funding for public education. It is not a coincidence that they chose to move to the street as a form of protest. When they take place in a piazza, simple gestures like carrying an old desk outside, and discussing with others a thought or a fact, inevitably spark the collective imaginary: is this how we should always run our schools? Can this be a model for how society – or at least a city – should work?

The symbolic meaning of holding classes in piazzas is a beautiful reminder of why public space is an important aspect of modern democracies. All over the world, people have taken to streets and piazzas to discuss pressing issues, to gather support, and to voice their consensus or disagreement. In Italy, in particular, open urban spaces have always been central stages and resources for historical, cultural, and political developments. The lives of public spaces, however, have always been fraught with contradictions and unintended consequences. This is because, if they can act as vehicles and embodied terrains for powerful imaginaries, the latter do not always refer to the same publics,[1] or to the same dreams. Another, very different example is illustrative here. In April 2007, a parking ticket given by a municipal officer to a Chinese-Italian woman sparked a "street war" in Milan, which included violent clashes between residents and police, the burning of cars, and injuries to more than a dozen people. During this event, three hundred demonstrators, mostly Chinese-Italians, rallied to protest against police and city authorities (see "Guerra di strada," 2007), sparking, in turn, resentful street meetings by right-wing political parties.

The incidents took place in the Canonica-Sarpi neighbourhood, an area often called the Chinatown of Milan – even though Chinese-Italians represent a minor portion of its residents[2] – due to the high concentration of wholesale clothing stores run by Chinese-Italian owners. Because the coming and going of trucks carrying merchandise often disrupts everyday life in these streets, and because these stores supposedly do not "look Italian," many people in the city have started to resent the presence of Chinese-Italians in what, they feel, had long been a traditional Milanese neighbourhood. In fact, starting in 2008, the municipality has carried through a series of interventions with the goal to give the area a more solidly "Italian" look, including the creation of a wide pedestrian area with cobblestones in one of its main streets.[3] At the heart of the April 2007 struggle, then, were opposing views on the identity of this area, and on who could contribute to it.

This conflict, and the arguments that followed, were neither a unique nor an isolated incident. Recurring and bitter disputes over who should occupy Italian city spaces and how uphold the central role of public space and reveal the discrimination that often accompanies its everyday use. Notions of culture, race, and ethnicity were key elements in this particular incident, but conflicts over spaces are intimately linked to a whole range of important issues that are shaping contemporary Milan. The people I met during my research in 2004–05, 2009, and 2011 worried about the cost of living and complained bitterly about the endemic corruption that was making the sense of general crisis worse. In markets, on buses, and in the halls of public meetings, they told me how changing demographics were affecting their everyday lives as they cared for both children and older parents, and they discussed the lack of affordable housing that was causing more and more people to settle away from the city in neighbouring municipalities; several stories were making the rounds about elderly residents on very meagre pensions who had to resort to eating pet food.

Milan has long been a city in transition. In the period after the Second World War it was a key industrial centre, with large working-class neighbourhoods and major factories. In the past three decades, however, Milan's shift from an industrial to a service-oriented economy has affected the social and spatial fabric of the city. Since the late 1980s, neoliberalization in Italy has eroded social assistance measures and made employment more temporary and uncertain. Processes of deindustrialization have been accompanied by immigration, an expanding multiculturalism, and more recently, the redesigning and redevelopment of extensive urban areas. Such changes have had an impact on the role accorded to memory. As neighbourhoods are gentrified, new ones constructed, and an increasingly diverse population uses the city's public spaces, the ways in which the inhabitants of Milan remember their city and streets can become a source of constructive belonging as much as a tool to exclude others from that community. These social circumstances and the shifting positions of people in the city then become visible in how sectors of the population deploy and seek to control public spaces, through everyday negotiations, citizens' actions, or even outright "street wars" (Ciorra, 2003; Maritano, 2004).

In this book I trace how people and social groups participate in public spaces in Milan.[4] My analysis grows out of a set of ethnographic questions. First, I ask how and why the presence of various publics in its piazzas and streets serves less to foster an appreciation for diversity and equality than to strengthen a sense that "others" are taking over the city. In contemporary Milan, in fact, who uses which public spaces, and how, can easily become a way to distinguish between people assumed to be entitled to the city and people who do not "really" belong there. At the same time, I show that these claims do not go unchallenged. Through stories, memories, daily journeys, and active appropriation of spaces, less-privileged inhabitants seek to redefine their roles and relationships to the city.

One of the aspects that makes Milan an interesting location for this inquiry is that the recent yet significant arrival of people from other countries has rendered its public spaces into particularly charged sites for the negotiation of inhabitants' places in the social, economic, and cultural landscapes of the city. These dynamics complicate and deepen existing urban inequalities and conflicts, at a time when neoliberal shifts are reorganizing social and spatial relations.[5]

Milan, moreover, harbours a vibrant culture of activism concerned with social justice and a long history of mobilization against dominant institutions and forces. Left-oriented oppositional Community or Social Centres (*centri sociali*, described in detail in chapter 4), for example, have been in existence since the 1970s, and these have involved several age groups and generations in their activities. Migrants' rights associations, oppositional architects and designers, and Christian-inspired social justice organizations are just a few of the many other activist groups working in the city. Such activist practices, and the significant socio-economic changes that have affected Milan in the past fifty years, make this metropolis into an especially intriguing locale for tracing diverse interpretations and claims to urban spaces.

Second, in this book I interrogate the very concept of public space and offer a reflection on how we can study it ethnographically. My aim is thus methodological, too: to propose strategies for research in large, complex cities. The struggles and negotiations I witnessed in Milan, in fact, pertain not only to the uses of public space, but also to the very meaning of the term. Simply put, when I was trying to explore people's experiences of and ideas about public space, I was faced with a category that was melting before my eyes. Whereas, when I started my research, I thought of public space as simply piazzas, parks, and streets that are not privately but publicly owned, and that are accessible to all, for the people I talked to there was no such thing as a clear definition of public space. People in Milan included all sorts of locations in this category, such as coffee shops, cinemas, bars, courtyards, public health centres, churches, subways and public transit, illegally occupied buildings, and *centri sociali*.

This ambivalence became central for my work. For one, my interlocutors suggested that public spaces include places that "work as such" for particularly situated social actors. Some of my interlocutors who had recently immigrated to Italy, for example, showed me that private spaces used by immigrant groups constitute important sites of sociality and serve as public spaces because they are less dominated by white, Italian-born Milanese than central open piazzas are. For another, my interlocutors emphasized that public spaces are tied to a whole array of ideas, practices, and histories, themselves actively debated in a social context. By this I do not only mean that the everyday lives of public spaces – who uses streets and piazzas, when, and how – reflect different inhabitants' positions in society, as well as different important political and economic structures. What I seek to foreground in this study is that, for all the people I met, talking about, journeying in, and using public space was a way to comment on other and all aspects of city life.

One of my interlocutors, for example, a young, low-income man in a temporary, uncertain job in the service sector, answered my question, "What are the public spaces that are important for you in the city and that you use in your everyday life?" by leading me on an extensive itinerary both on foot and using public transportation. In this journey, he presented the public spaces of the city as being a network of interconnected sites, with our movements through them being strikingly similar to the activity of surfing websites. Our tour together became both a description of Milan and a reflection on the role of the Internet in his life. Other interviewees used public space as a stepping stone to talk about the courage and work ethics of foreign newcomers, to map commuter routes in and out of the city, to offer observations on the importance of water, and to reflect on urban redevelopment projects. At first I was puzzled over these different reactions; during my fieldwork, however, I learned that rather than being a limitation or a disadvantage, the very slipperiness of public space – its capacity to conjure up and evoke divergent conversations – was part of what made it interesting and an active force in Milanese society.

Following my interlocutors' insights, I argue that public space is a fleeting concept and an ever-changing, shifting construction. On such uncertain terrain, people both negotiate what public space is, or can be, and seek to stake their claims to it. Rather than trying to define what public space is, in this book therefore I am interested in the interplay between its different meanings and enactments – and consequently, in how public space can serve as an ethnographic point of departure for tracing local debates and concerns. By attending to the particular ways in which public space becomes significant in Milan, my work wants to connect more extraordinary moments like the open university or street demonstrations with far more ordinary, daily, and regular engagements with urban spaces. I suggest that although public space is a crucial locale for political campaigns, rallies, and revolutions, it is often through more subtle and intimate relationships with spaces – and with the imaginative, discursive, and performative possibilities they engender – that city users come to grips with complex social categories, local history, and political dilemmas.

Importantly, I argue that in Milan, a significant part of these engagements occurs in a visual form. My interlocutors showed me that to see and be seen is an integral part of contributing to urban life and recognizing others who share the same city spaces. One of the people I met in Milan in January 2009, for example, Riva, a young woman who had migrated to Italy from North Africa and who was working as a nanny, explained that in the city everyone judges by appearance and that because of this it would be impossible for her to wear her "African clothes." If she did that, it would be even harder for her to find a job, as she would become even more vulnerable to people's judgments and negative preconceptions against immigrants. On the other hand, explained Riva, cultivating style and aesthetics is a way for her to *be* Milanese as it enables her to take part in a city in which people use aesthetics as a pervasive social idiom.

Riva's comments are illuminating in more than one respect. First, her experience demonstrates that vision is never innocent, transparent, or straightforward. Rather, it is a productive force that participates in the very creation of the social. The staging of cities, for example, and the careful managing of their landscapes, can work to strengthen political regimes, uphold unequal social relations, convey ideas, or give weight to particular claims (Rademacher, 2008; Potuoğlu-Cook, 2006; Blomley, 1998). Second, Riva's description shows that seeing is a social practice that happens in specific places and with specific audiences and that helps position both the seer and the seen in a social context. These practices of vision are something we continuously learn in particular social and cultural contexts, thus constituting a "form of social apprenticeship" (Grasseni, 2011: 21). Echoing Pink's work (2007), Grasseni recommends there be more ethnographic studies that examine practices and cultures of seeing and their connected social and political contexts.

Milan is an especially fertile ground for researching the role of vision as a social practice, because in this city aesthetics is at the basis of a complex cultural, economic, and social system. As fashion and design grew in importance during the deindustrialization of Milan, people's attention to "how things and persons look" has resulted in a multifaceted and intricate relationship between fashion and race, gender, and class that necessarily involves public spaces as sites for the display and negotiation of social difference. What I argue in this study is that in Milan the very constitution of particular publics is understood and brought forth also in a visual idiom, which helps organize the spaces inhabited by these subjects. Public spaces thus emerge as complex fields of vision, shaped by and shaping social relations also through "struggles occurring at the level of the visual" (Pinney, 2004: 8).

The events that took place in the week of 22 April 2009 are a telling example of some of these aspects. The annual Design and Furniture Week was bringing to Milan large numbers of architects, designers, interior decorators, furniture producers, and professionals, as well as students associated with those fields. Cocktail receptions, exhibitions, and demonstrations in several stores throughout the city were contributing to an atmosphere of effervescent excitement and busy attention. On opening night, I joined the crowd that was circulating from venue to venue and stopped at the display of a transparent glass kitchen in a transparent room that had been stationed on the sidewalk, just to the side of one of the oldest churches in Milan.

The ancient bricks of the church wall were shining through the glass of the kitchen stove, shelves, and counters on display, creating a surprising effect and highlighting the aesthetic pleasures of things both sacred and mundane for the discerning crowd. A few moments later, the piazza onto which the church faced was filled with people gathering for a large rally, shouting slogans in support of immigrant workers and denouncing the difficult housing situation in Milan. The rally interrupted the circulation of the design crowd, which stopped to listen to what was going on.

This moment not only crystallized the different social positions, identities, and interests of two different groups of people who were using the piazza that night. It was also a coming together of two different kinds of spectacle: on one side, placards, flags, and the powerful display of the mass of people on the piazza; on the other side, the exhibition of a carefully cultivated creative aesthetic. I do not want to indicate that one was better or more important than the other, nor that they are necessarily always opposed to each other, or even associated with these particular subjects. (The comments of Riva above remind us that one can be both an immigrant and interested or involved in fashion and design; and many of the designers and architects are very sympathetic to migrants' rights.) What I seek to emphasize is that they both play a significant role in Milan not *in spite* of being visual representations, but *because* of it.

The sets of images and the ways of seeing and being seen deployed that night were not simply reflections, or exterior expressions, of systems and structures. They were themselves ways to engage with those social forces. The rally for migrants' rights was a way to make immigrants visible as a public presence in order to counter the social marginalization that many of them experience in their lives, while recognizing, using, and producing style and aesthetically pleasurable objects is a practice that helps consolidate the social, cultural, and economic role of many middle-class professionals in the city.

Recognizing the rally as a provocation precisely because it was directed at the particular aesthetic displayed by the design event, newspaper reports the following day (23 April) complained that the immigrant rally was damaging the image of Milan at a crucial moment in which it was showing itself off as a cosmopolitan and fashionable setting for style on an international level. In a similar vein, we can see the rallies, posters, and stencils on city walls of the independent and oppositional Community Centres, allies of immigrants' rights, as disturbing the neoliberal city, not only by protesting urban inequality and precarious working conditions, but also by tampering with the very appearances of the post-industrial metropolis.[6]

Through my attention to vision as a social practice, this book foregrounds performative engagements in and through city locales. In my fieldwork, this aspect emerged both during particular events and gatherings (like in the example above) and in far more ordinary everyday life, indicating that people's engagements with public spaces are always steeped in performance. People fashion a "sense of place" through the extensive repetition of acts of remembering, looking, telling, and moving through the city (Guano, 2003; see also Guano, 2007; and Fabian, 1990). In turn, inhabitants enact, express, and negotiate their social identities, links to a community, collective memory, gender roles, and political orientations (to name just a few) in and through space.

These understandings of self and other are not simply pre-existing their expressions in public locales. As Fleetwood describes, "identity is not a state of being but one of doing" (2004: 40). Identities are not fixed and stable essences, existing before and apart from society and culture. A focus on performance helps us pay attention to how identities are constituted through social interactions and practices, some more ephemeral than others. Most importantly, it is through situated and embodied performances that particular constructs like public space emerge. Social action, by literally taking place in the city, shapes and generates particular meanings of space. A performative approach to urban life thus converges with an understanding of space not as a predefined locale with particular functions, but as a process unfolding in time, something "done by people" (Weszkalnys, 2010: 17), and appropriated as a lived dimension in accordance with and against the grain of dominant social relations and structures (Schielke, 2008; de Certeau, 1984; Lefebvre, 1991). Moreover, it reminds us that the memories and histories that are linked to specific sites are called upon and interpreted from the vantage point of the present and through particularly situated storytelling practices (Roseman, 1996).

Although this holds true for urban space in general, it is even more relevant for piazzas and streets because they are key locales where social and political identities are negotiated and made visible to others. As a number of scholars have suggested, performative practices in public spaces are central for the establishment, maintenance, and strengthening of dominant powers, as well as for the emergence of progressive social movements (Guano, 2002; Taylor, 1997). In relation to the latter, public space allows new claims and movements to manifest themselves in the social sphere and to acquire a social body (Mitchell, 1995).

Urban spectacles and theatricality, whether emerging through extra-ordinary moments or in subtle everyday forms of living, can serve as a "modality for the public negotiation of citizenship" (Guano, 2002: 306; Goldstein, 2010). The latter should be understood not only as a set of norms or an institutional category, but also as a relational, embodied, and constantly constructed aspect of people's lives and their social positions (Fikes, 2009; Partridge, 2008). These considerations are relevant for tracing the daily manufacturing of a sense of belonging to the city – here understood as an affective, imaginative, and material link constituted both through self-identification and recognition by others (Jackson, 2005). Indeed, in Milan, everyday conversations and practices of relating, attunement, and distancing between inhabitants also partake in wider discourses concerning who is a "Milanese" and what the Milanese share. This excerpt from my fieldnotes is telling in this respect:

On a bus, there is an elderly woman looking for somebody to chat with. She tells me she moved to Milan from Puglia, a region in the South of Italy, when she was 18 years old. "[It is as if] I was born in Anfossi street!" she explains emphatically, to show me her attachment to the Milanese neighbourhood she has been living in almost all her adult life. Another woman on the bus overhears our conversation. She turns towards us and comments angrily: "Then you are NOT Milanese! You were not born here!"[7]

That same afternoon, I enter a bakery with my son. An older Egyptian-Italian couple runs it. The man comes and goes from the back of the shop, bringing croissants, focacce, and buns. The woman wraps my bread in a white paper bag. It's six o'clock, the city is busy; in the small quiet shop we are all looking tired. The two bakers stop and look at me for a while. Perhaps they are trying to guess who my son's father might be, seeing that he has darker skin than I do [they will ask me about him the next time I pass by the bakery].

"Where do you come from?" they inquire. I am surprised. It is usually "Italians" who ask "Egyptians" where they are from. "From Milan," I say. "I was born here." "Ah, but really Milanese?" They ask with evident surprise. I am looking for words. How to say that I might be Milanese in some ways but not in others? That they are perhaps more Milanese than me? That still I am aware that many people in Milan would consider me to be Milanese but not them? And would they think this all matters? "Yes," I respond, "I am Milanese" and quickly add, "but I do not live here. I live in Canada right now. But yes I was born here, my parents and brother and cousins live here, and I feel at home here." (26 Nov. 2004)

As these examples illustrate, the word "Milanese" is less a fixed category than a powerful referent that can be used in many different ways. It usually indicates people born in Milan, but depending on the speaker, it can also include those who have lived in the city for a number of years. Often, "Milanese" is used in an exclusionary way, to mark degrees of entitlement and belonging to the city and to discriminate against immigrants from other countries and migrants from southern Italy. In this book, unless otherwise indicated, I use "Milanese" in an inclusive way, referring to all people who are currently residing in Milan.

On the opposite end of the spectrum, the word *extracomunitario* designates a person who has come to Italy from outside the European Union. Commonly used to refer to immigrants, this term means literally, a "person who is not part of the community." A less discriminatory designation for immigrants is foreigner (*straniero*, which also means stranger). This word, when used by institutions and in statistics, refers to residents who have non-Italian citizenship. Everyday perceptions and negotiations, however, complicate this definition. "Italian-ness" is commonly associated with a Christian orientation, is recognized in particular cultural practices, and is racialized as white, rather than being equated with formal citizenship and/or residency in the country. For this reason, "Italians" rarely consider that the "immigrants" they encounter could actually be Italian citizens just like themselves; to remind the reader of these complexities, in this book I sometimes use hyphenated identity appellations such as Egyptian-Italian, although people in Milan do not generally use these terms.

A visible minority resident is thus almost automatically categorized as an "immigrant" with no consideration of his or her nationality or of the length of time he or she has lived in Italy. Black inhabitants are particularly singled out as foreigners, and the most common picture of an extracomunitario is that of a Black, single, young immigrant man from Africa. Exclusive understandings of residency and entitlement to the city are challenged in everyday life through casual commentaries, organized projects, and political action. More important for my point, everyday negotiations of these categories, like the ones in the anecdotes above, can be fruitfully seen as performative enactments. Paying attention to them as performances highlights the contextual creation of social categories – and the way they interpellate, add to, or interrupt wider ideas and discourses – as well as the specific locales where they take place.

In relation to the latter, while public spaces can become significant stages for the performance of self and others, and for voicing one's claims in relation to the state and the wider public sphere, what I found equally compelling in my research was the inverse yet concomitant process: the way in which performance rooted people in public space and made particular meanings of the latter intelligible. This emerged for me most clearly during the walking tours that my interlocutors guided me on (and that I discuss in more detail in the chapters that follow). As my various guides performed being certain kinds of inhabitants, their particularly enacted "Milanese-ness" tied them to streets and piazzas in specific ways. In this process, public space as a category acquired particular shapes and meanings and harboured specific openings for discourses, imaginaries, and actions.

Milanese Encounters: Doing Fieldwork in Italy

To investigate public space and the role of vision in Milan, I carried out ethnographic fieldwork in the city from October 2004 to April 2005, from January to May 2009, and in May and June of 2011. My work also profits from and incorporates my prior knowledge of the city, where I was born and where I lived for over twenty years. Within my fieldwork periods, the very end of 2004 and the beginning of 2005 was marked by a series of encounters with people who shaped in crucial ways my thinking about urban space and vision. The insights of these interlocutors – Don Felice, Mohamed Ba, Francesca, Maria Anacleta, the Schuster Youth, Giacomo, and Alice – constitute the core of the book.

The latter two persons are part of the movement of the Italian Social Centres that I describe in chapter 4, a loose network of oppositional community centres that are created by "occupying" unused buildings. In this book, I discuss three of these Centres: Leoncavallo, Casa Loca, and the Stecca degli Artigiani. Other organizations or groups that were particularly inspiring for my research were Terre di Mezzo, Naga, VivereMilano, and a group of students who participated in a workshop called "Building Zenobia," housed in Casa Loca. The list of participants in Table 1 provides a summary of these key interlocutors.

In terms of urban space, my 2004–05 research represents a particular period. Although there were already several very important redevelopment projects underway in the Milanese territory in 2004–08 (e.g., in Bicocca and Portello), urban renewal had not yet altered dramatically the landscape of Milan. I found that, generally speaking, it was a moment of relative stability and calm, in which long-standing relations between spatial elements and key dilemmas of public space and visuality came

Table 1. Main Interlocutors/Guides and Organizations/Networks That Appear in the Book

		Year of Encounter and Location in the Book	
Eliza	A working-class, elderly, white, Italian-born woman; my guide for a walking tour through the Bovisa neighbourhood	2009, 2011	Chapter 2
Alice	A working-class, elderly, white, Italian-born woman; one of the Mothers of the Leoncavallo, an association of women involved in the Centre for more than two decades	2004, 2005	Chapter 4
Giacomo	A young, white, Italian-born student; one of the founders of Casa Loca	2005	Chapter 4
Schuster Youth	Two young women students who migrated to Milan from El Salvador in their teens; participants in the Schuster Centre church activities	2005	Chapter 5
Don Felice	A young priest and student who migrated to Milan from Mexico; one of my walking tour guides	2004, 2005, 2009, 2011	Chapter 5
Francesca	A middle-class, middle-aged, white, Italian-born woman; one of my walking tour guides	2004, 2005	Chapter 7
Maria Anacleta	A middle-aged woman who migrated to Italy from the Philippines; one of my walking tour guides	2004, 2005, 2009	Chapter 7
Mohamed Ba	A middle-aged, Black, community educator and theatre writer who migrated to Milan from Senegal; one of my walking tour guides	2005	Chapter 8
VivereMilano	A movement started in 2004 by 30- to 40-year-old residents of Milan; members participate in a blog and in urban initiatives to discuss and seek solutions to what they perceive are key problems in Milan	2005	Chapter 3
Leoncavallo	The oldest, best-known, and largest Social Centre in Italy; started in 1975, it occupies a large disused printing factory in the northwest of the city; see www.leoncavallo.org	2004, 2005	Chapter 4

(Continued)

Table 1. (Continued)

		Year of Encounter and Location in the Book	
Casa Loca	One of the youngest Social Centres in Milan; started in 2003, it is located in the Bicocca neighbourhood and occupies a former Pirelli factory workers' building; see www.casaloca.it	2005, 2009	Chapter 4
Stecca degli Artigiani	A Social Centre in the Isola neighbourhood, active since the 1980s; previously located in an "occupied" former factory, after the latter was demolished in 2007 as part of the redevelopment of the area, the Stecca moved to a temporary home close to its former location.	2005, 2011	Chapter 6 and Conclusions
Naga	An organization promoting the rights of marginalized inhabitants and offering free medical care to immigrants, illegal residents, Roma and Sinti, and refugees; see www.naga.it	2004, 2005, 2011	Chapter 6
Terre di Mezzo	Both a street newspaper and a publisher of books on Italian and international social issues; see www.terre.it	2004, 2005, 2009	Chapter 6
"Building Zenobia"	A workshop held at Casa Loca in March 2005; architecture students, professors, and activists discussed vacant industrial areas and their possible uses	2005	Chapter 6

more easily into view. Conversations with people during 2004 and 2005, then, highlighted processes and dynamics that continue to be central and, indeed, have grown in importance in the past few years and are thus crucial in understanding very recent transformations.

The research periods of 2009 and 2011 allowed me to observe subsequent developments, events, and initiatives and provided me with another group of interlocutors and an additional body of conversations and itineraries. I also re-established contacts with some of my earlier participants, including Don Felice, Maria Anacleta, and Marta and continued our conversations on the topics of public space and visuality. This period, moreover, marked a series of important transformations in the Milanese landscape. In 2011, especially, Milan seemed to be at a crossroads. The elections, the economic crisis, and the visible changes of several neighbourhoods increased the sense of uncertainty felt by many residents. Many people from middle-class backgrounds were hoping to leave Milan. As Don Felice explained in a conversation of 2011,

> I seem to be the only one wanting to stay. Everyone else is trying to escape.
> (7 June 2011)

For these reasons, I chose to juxtapose stories, encounters, and events from all three periods of research, while using the insights of my key 2004–05 informants as points of departure for my analysis.

All three fieldwork periods entailed participant observation of places and events, as well as unstructured conversations and in-depth interviews[8] (or series of interviews) with a total of forty-five people. Participants were drawn in one of two ways. About half were the result of personal connections and/or of snowball sampling. The other half of my interlocutors and guides were people I met in various urban places – mostly in one or another of the piazzas and streets of the centre, but also in parks, aboard buses, in Social Centres, in open courtyards, or on university campuses. Except for Mohamed Ba, all names used in this book are pseudonyms. Mohamed Ba wanted me to use his name, and I honoured that request, also to emphasize his role as author of the wonderful theatre representation of the history of Milan (that I present in chapter 8), a public speech that he has delivered for the local street newspaper *Terre di Mezzo* in recent years.

Here I want to point out that as a person who was born and grew up in Milan, I could most times be seen as an "insider" to the city. However, as my trip to the bakery in my earlier anecdote shows, my peculiar position as a subject in-between-places muddled this status. As a native and a stranger, an Italian and a Canadian, a low-income student (in 2004–05), and a member of a middle-class family, a married woman with a husband and two children and yet a single mother in Italy (in 2004–05 my three-year-old son was with me in Milan, and in 2009 I was accompanied by my two-year-old daughter), I was often at least two things at once.

For example, in relation to international migrants like Maria Anacleta, Marta, Riva, the Schuster Youth, and others, I was both part of a society that often discriminated against them and an immigrant from somewhere else. On the one hand, I too had a divided family and lived the intense longing for a home that most of the time was far away. Many of them asked me for information about life in Canada, visas, and how it is to be there as an immigrant. On the other hand, this all did not take away the fact that I could pass as a "regular Milanese" in most circumstances. For journalists and researchers from *Terre di Mezzo*, a street newspaper, I was a kind of journalist, too, interested like them in stories, spaces, and issues of social justice. One of the main differences between us, however, was that we were writing for different audiences. While the people from *Terre di Mezzo* are part of a project to educate residents of Milan about poverty and inequality, I was writing mainly for academic readers.

When I visited Social Centres I could be an ally and a sympathizer, and I brought with me a history of involvement with activist groups in Germany and Canada. However, not having been part of local and strenuous occupations and actions clearly made me an outsider to their project. For white, Italian-born Milanese women I was both somebody they could relate to and recognize as "Milanese" and a "strange native" who had forgotten some of the ways to be in Milan through my recurrent, long absences. Because they saw me as needing their guidance and instructions, they often positioned themselves as teachers.

In relation to VivereMilano (the association I talk about in detail in chapter 3), as a (temporary) resident in my thirties, I found myself an automatic member. I appreciated the enthusiasm of the members of this association and their wanting to go out into the streets to do something good about the city, but I felt uncomfortable being counted in; thus, while I had been coincidentally wearing an orange scarf all through the autumn of 2004 (orange being the colour chosen by the group to represent itself), after December I often left it at home or hid it in my pocket.

Exactly the opposite happened, however, in 2011, when the political coalition against Berlusconi and its local representative Moratti decided to use this same colour. Ironically, while I resurrected my orange scarf and sympathized with the coalition, I found myself excluded from the actual electoral process, as I was not entitled to vote even though I am an Italian citizen, because I am not a permanent resident in Milan. These complex positions resulting from doing research "at home," in which I was both an insider and outsider in different ways in different circumstances, alerted me to wider politics of identity and difference in contemporary Milan, urging me to pay attention to the contextual and dynamic creation of both public space and culture within and across particular urban locales.

Last, but not least, the map of Milan (Figure 1) shows the most significant places I refer to in my discussions. My research included key

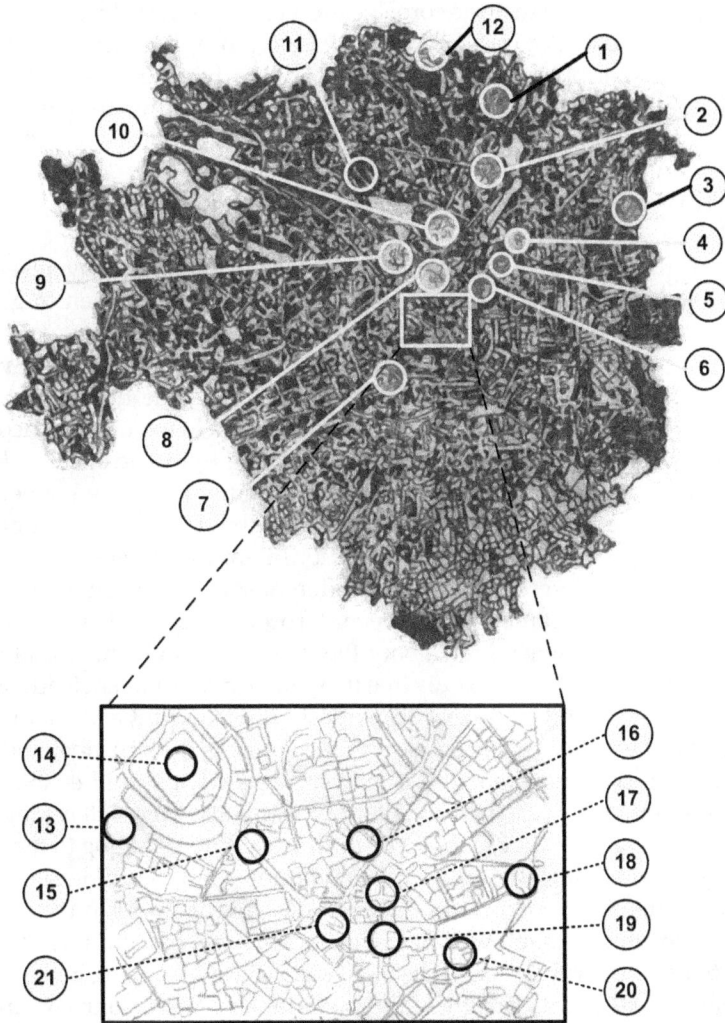

Figure 1. Map of Milan. The numbers in the map refer to the following places (all locations approximate): 1 Casa Loca and Bicocca; 2 Leoncavallo; 3 Schuster Centre; 4 Piazzale Loreto/Loreto Piazza; also the peak of Moroni's "North Triangle"; 5 Corso Buenos Aires; 6 Porta Venezia; 7 Navigli area; also the peak of Moroni's "South Triangle"; 8 Brera neighbourhood; 9 Canonica-Sarpi; 10 Isola neighbourhood, Stecca degli Artigiani, and Garibaldi Repubblica; 11 Bovisa; 12 Zenobia's dismissed area; 13 Cadorna Station; 14 Sforza Castle; 15 Via Dante/Dante Street; 16 La Scala Theatre; 17 Galleria Vittorio Emanuele; 18 Piazza San Babila/San Babila Piazza; 19 Piazza del Duomo/Duomo Piazza; 20 Piazza Fontana/Fontana Piazza; 21 Piazza Mercanti/Mercanti Piazza.

public spaces in the historic core; locations such as the Schuster Centre, the Leoncavallo, Bovisa, Bicocca, and Casa Loca, that are close to or part of the periphery; and the Isola neighbourhood that is quite centrally located, but could still qualify as a periphery within the city centre (Zajczyk, 2005).

Walking in the City[9]

One of the strategies that proved most fruitful during my research was guided city walks: I asked eight women and three men to guide me through "their" Milan to show me the public spaces that were significant for them. During these itineraries, my guides interacted with people and places around us, telling me at the same time about the city and about their lives. As Richardson (2008), Guano (2003), Nicolini (1998), and Pratt (1988) show, a simple walk in the city can elicit a whole universe of stories, memories, and interpretations, deeply connected not just to one person's experiences but also to wider social, economic, and cultural processes and structures. It did not surprise me, then, that my interlocutors were interested in different sites and, even when they brought me to the same locales, they interpreted them in distinct ways.

Here, however, when I talk about the diversity of the itineraries I was guided on, I mean to say not only that they showed me different landmarks and associated different memories with those places, but also that these itineraries became a practice aimed at recognizing, constructing, and/or articulating particular forms of public space as what would enable our very encounter. To express it differently, these walks were interesting for my project, because they embodied specific ways of belonging to the city, thus proposing different perspectives and frameworks for understanding Milan. Similarly to what Sarah Pink describes for her tours in Wales, my companionship through walking became a practice of "attuning" (2008: 175) to my interlocutors' embodied and sensual ways of being in the city.

Pink (2008) points out that itineraries are a shared journey, and I found that my presence clearly shaped what my guides decided to show me. Indeed, the guided tours were akin to what Fabian (1990) and Castañeda (2006) describe when they say that an ethnographic inquiry is a kind of collaborative theatrical endeavour. As Fabian argues about ethnographic practice more generally, the tours were less the obtaining of particular information and more "occasions" (1990: 7) for creating knowledge with others through cultural enactment and situated representations. Much of culture and social action, in fact, cannot "simply be called up and expressed in discursive statements" but can be apprehended only through cultural performances (6). As an ethnographic form of inquiry, moreover, it was the open-endedness of walking tours that worked well for my participants and me. If public space in Milan is important as a point of departure for many different conversations, guided itineraries were helpful because they allowed this diversity to emerge. For this reason, guided itineraries were instrumental for my focus on encounters as a practice of knowledge creation.

The very concept of the itinerary moreover became a useful model for my fieldwork and for people's connections to their city. First, I started to regard my research itself as a lived itinerary through Milan. Not only because, like the itineraries of my guides, it links places and stories, things forgotten, remembered, and longed for, imagination, and curiosity. Confronted with the dilemma of how to do ethnographic fieldwork in a city obviously too vast for it, yet not wanting to study only a particular neighbourhood, I learned from the suggestions of my interlocutors and followed their ideas along and across different discourses circulating in the city. My work then became a multi-sited ethnography within Milan, in which my daily life and research questions were the embodied journey linking places, people, practices, and social structures. To use Calvino's words, I found myself moving in a forest of tales, a wood that "is so thick that it doesn't allow light to pass [...] where you can move in all directions, as in space, always finding stories that cannot be told until other stories are told first, and so, setting out from any moment or place, you encounter always the same density of material to be told" (1981: 109). Just like for Calvino's forest, it became less important to grasp the city in its entirety and more to be moved and move along its lifelines – to trace how each story was connected to another, whether buried within it, running slightly askew, or forecasting its telling.

When talking about a multi-sited analysis, I seek to emphasize the incongruent positions of these sites vis-à-vis each other and the city terrain. The different angles I adopted, and the specific situations and locales that enabled them, are not to be understood as knowledges and points of view that can stand side by side and compose a fuller picture (see also Tsing, 2005). Rather, these perspectives – or entry points into the social made possible by the shifting roles of public space itself – often constitute different modes of learning, knowing, and moving through Milan. In contrast to Pink's use of a walking methodology, in fact, I found the difference – sometimes even incommensurability – between itineraries to be the analytical centrepiece of my work.

Second, I started to look at people's relationship with the city as a nexus of physical, social, and metaphorical journeys. Appreciating the complexity of the latter helped me to understand how both connections and disconnections between differently positioned people shape their place in the city and in the wider society. Reflecting on who can meet in which public spaces, who does not, and in which contexts is a useful way to examine how power and social difference shapes public space and different subjectivities. It greatly struck me, for example, that Italian middle-class women rarely interacted in public space with low-income migrants, even if the latter were often intimately involved in their domestic lives as nannies and caregivers. Urban tours were helpful in this respect, as they helped me to trace people's itineraries in a social landscape and to reflect on pathways of avoidance.

Let me emphasize here that, while walking with informants has gained a significant prominence in the literature in the past few years (see, e.g., Ingold and Vergunst, 2008), when I started to use walking tours in 2004, my initial reason for doing so was that they grew out of the distinctly local ways of relating and being in the city of Milan. Going for a walk is a preferred way for many Milanese, especially women, to engage in conversations and relate to city spaces. Walking is thus, similarly to what Richardson (2008) documents for Odessa, part of a familiar repertoire of inhabiting the city. As a research practice, walking tours made sense to my interlocutors because they were *already* a local methodology – ways for people to make sense of Milan and to gain insights about themselves and others.

Importantly, what made walking a fruitful methodological strategy was that it is always a socially embedded activity and one deeply entangled with visual cultures. We could say that in Milan (and in urban Italy more generally), people never walk alone, even if they are unaccompanied by others. Their steps and routes always respond to dominant discourses and ways of being and to oppositional forms of inhabiting the city. Guano (2007), for example, describes how women, in the northern Italian city of Genoa would walk as a way of spending time outside as it is not proper for them to simply "hang out" in open places. Also, as many of my interlocutors suggested, walking is an important way to enjoy the city for people who have little disposable income.

One of the striking aspects of itineraries was that they invited imaginations, "ghosts," and invisible cities (Carter, 1997) to enter walkers' experiences. These moments, rather than being divorced from cultural practices, strongly situated residents in relation to social processes and structures, as they became reflections on historical changes and on contemporary debates of possible futures. As my guides reminded me, with every footstep we retrace the city that we are living in and the ones that we are dreaming about. Last, but not least, for many of my guides, walking was an opportunity for learning. Using walking tours as a methodology, in this book I seek to contribute to scholarly attention to walking as a mode of attaining knowledge (Ingold and Vergunst, 2008) by emphasizing its visual components and deploying it to research a metropolis shaped by deindustrialization and neoliberalism.

Organization of the Book

Spanning the period from 2004 to 2011, each chapter of this ethnography (with the exception of the introduction and chapter 1, "Orientations") is structured around an encounter – or set of encounters – that taught me specific ways of understanding and interpreting Milan.

The initial section of the book introduces the reader to the key concepts of the study. In "Orientations" (chapter 1), I chart an interpretive framework for researching public space as a slippery yet productive object of analysis and discuss vision as a social practice that is embodied, multiple, and conflictual. In chapter 2, "Milan," I describe the city to the reader, while elaborating on the relationship between ethnography, history, knowledge, and voice.

In chapters 3 to 5, I describe different ways of enacting and understanding public space in Milan. In chapter 3, "The Agora of the City," I examine how a street forum organized by the largely white, Italian, middle-class association VivereMilano uses the Duomo Piazza in a grassroots effort to improve city living, all while marginalizing issues of importance to less-privileged residents. I compare this event with a 2011 electoral gathering to highlight the role of hope in the lives of public spaces. In chapter 4, "Spatial Politics," I discuss how Social Centres depict the urban territory as deeply political and as a central locus and object of struggle. Their theoretical perspectives of Social Centres and their actions are rooted in an alternative historiography of urban space, consciously opposed to more dominant understandings. Finally, in chapter 5, "Creating Spaces, Constructing Selves," I look at some aspects of the relationship between immigration and city space. Although public spaces are a precious resource for several of the new immigrants that we meet here, immigrants are usually the ones who have the most difficulty in claiming these spaces as their own. In this chapter, I argue that comments about public space in Milan often tend to essentialize culture. In everyday comments, in the media, and in common discourse, culture is seen as a variable that dictates the ways one perceives and uses urban spaces, rather than as something emerging and existing in social practice.

In chapter 6, "Entangled (In)visibilities," I examine how certain spaces, structures, and processes enable or produce (in)visibilities. My analysis in this chapter is informed by the insights of several activist groups that seek to draw attention to inequality in Milan. These groups' commentaries on Milan's empty post-industrial terrain are helpful for understanding both the spatial consequences of deindustrialization and invisibility as a constructed social category.

In chapters 7, 8, and 9, I describe three walking tours through Milan guided by differently positioned city residents. First, in chapter 7, "Walking with Women: Vision and Gender in the City," I compare the journeys of Francesca, an Italian middle-class woman, to the one of Maria Anacleta, who moved to Milan from the Philippines. For my two guides, vision itself becomes a way in which they relate to the city and find a place in it. A sense of belonging, albeit very different for the two women, emerged both when Maria Anacleta asked me to take pictures of herself in particular places to show her Milanese life to her family and through Francesca's talk about "beauty" in the city and her practice of seeing and recognizing it as she walked in its midst.

The differences and the connections between the women's itineraries show that modalities of seeing and appearing can often reinforce gender, class, and race hierarchies in urban locales. In turn, as I argue in chapter 8, "Stages and Places: Vision and Performance in Public Space," less-privileged inhabitants of Milan can use alternative ways of representing the city and their presence in it to help create spaces that would be more inclusive and egalitarian. In this chapter, by presenting a walk I was guided in by Mohamed Ba, I reflect on the relationship between public space, vision, and performativity. Believing that "theatre is an itinerary in hope," Mohamed Ba's commentary invites both the speaker and the audience to imagine the city differently and to welcome alternative voices and perspectives.

In the conclusion, "Into the Future," I describe parts of a journey through the Isola and the Garibaldi Repubblica neighbourhood as represented by a promotional video and by my own travel through this part of town. I take these different engagements with city spaces as points of departure to discuss the recent processes of urban renewal in Milan.

I argue that construction zones are monuments to the future, symbols of transformations and cosmopolitanism, and tools for the imagination. They are both the building grounds for new public spaces and key sites for local debates on the identity of Milan and its inhabitants. Here, too, an attention to public space and visual practices can provide helpful perspectives on contemporary Milanese realities.

Although my research focuses specifically on Milan, in this book I suggest theoretical and methodological approaches useful for the analysis of other cities as well. For one, the issues that are confronting contemporary Milan – such as gentrification, neoliberalization, social inequality, and struggles about the uses of streets and piazzas – are affecting many metropolises worldwide. These are, in fact, instances of significant trends, notions, and practices that have been occurring internationally (Paone, 2008; Drieskens et al., 2007; Smart, 2003; Caldeira, 2000) and are reshaping people's experiences of living in urban communities in places as different as Berkeley, Buenos Aires, and Berlin.

The ways in which these issues are experienced in Milan, and their effects on the city, can help us see linkages, value their effects, and formulate questions. For example, in what ways are conflicts over spaces in an age of neoliberal reorganization lived as visual struggles in different cities, and as such, how do they intersect with local, complex practices of remembering and storytelling? How are performative interventions part of efforts to reframe urban pasts and anticipated futures? How do people's embodied, affective, and sensual engagements with urban spaces respond to, challenge, and/or contribute to a politics of "containment" (Isin and Rygiel, 2007) and segregation that affect marginalized residents' positions and mobilities within urban territories? The discussions in this book participate in wider conversations and efforts by scholars, activists, and regular residents addressing the conditions that make cities sustainable, enjoyable, and just.

More specifically, this book seeks to enrich debates on Italian neoliberal cities – as analysed, e.g., by Muehlebach (2012) and Molé (2012) – by discussing the visual lives of spaces and how these participate in shaping unequal power relations between inhabitants. This analysis is grounded in visual anthropologists' insight that vision is a complex social practice with political consequences (Pink, 2007; Banks and Morphy, 1997). By researching a city where visual practices are central to urban engagements and to the constitution of public space, my work presents a case study for scholarly inquiries on cultures of seeing.

This book foregrounds the spatial dimension of debates and experiences of immigration and shows how discussions and engagements with public space act as a pivot for exclusion and discrimination, as well as for counterarguments to powerful discourses and counterenactments of spaces. My work echoes several anthropologists' ethnographic attention to the everyday negotiations of social categories in urban locales – such as Doninelli (2010), Fikes (2009), Partridge (2008), and Dines (2002); my goal here, however, is to put these processes in relation to wider discourses and shifting meanings of public space in the city, as it is their connections and incongruities that in important ways shape life in Milan.

For the other, and most importantly, this book shows that ethnography is uniquely placed in learning about the connections and disconnections that contemporary urban processes generate. My work suggests strategies for ethnographers wanting to explore public space and visibilities in urban communities. This "toolbox" includes walking tours as a research practice, performative-inspired forms of representing people's words, and a strategy of looking for connections between often fleeting and fragmentary moments of urban life. These ideas are suggestions for how to do research in cities, guided by a strong sense of anthropology and ethnography as a "sensibility" and as an "improvisational practice" rather than as a collection of instruments or notions (Malkki, 2007). As such, they leave room for following concepts like public space, which are not settled and fixed, but are best approached "obliquely,"[10] by taking their very indeterminacy as one of their key components.

What these strategies have in common, moreover, is that they are ways to learn from city dwellers as engaged experts and generators of frameworks, definitions, and perspectives. Both Steedly (1993) and Tsing (1993) criticize how anthropologists often see participants in ethnographies as merely providing examples for already established and supposedly universal concepts that then can be applied to local realities. Instead, the people and groups I met in Milan are for me the *creators* of theories on the city, visuality, and public space. They taught me particular ways of looking at public space and provided me with models to follow. Engagement with differently positioned inhabitants as co-creators of knowledge and critiques is helpful because it does not simply generate answers, but most importantly, helps us reconsider the very questions we ask. It can thus be a point of departure for an ethnographic practice that is continuously transformed through the encounters from which it emerges.

Chapter One

Orientations

At a bus stop in Piazza San Babila, a middle-aged woman waiting with me starts a conversation. I tell her that I am here in Milan to research public space. She looks at me inquisitively: "But do you mean spaces for gathering or actual piazzas and streets? Are those the same or two different things?" (Fieldnote, 12 Nov. 2004)

At the time, I had just started fieldwork in the city. The woman's questions left me wondering: what exactly is public space? Is it even a particular "thing" we can study? How do we recognize it? And how might a definition itself limit the way we think about the city and the ways people inhabit it? To begin unravelling some of these complexities of public space, it is helpful to briefly consider some of the ways in which I encountered it as an object in the field.

In Milan streets and piazzas are busy and very dynamic sites. Each of the piazzas of the Milanese centre has been used for decades, sometimes even centuries, by many (and sometimes antagonistic) groups and individuals to engage in widely different activities – such as setting up markets, walking with friends, holding religious processions, participating in armed battles, and much more. The San Babila Piazza, for example – where the rally for immigrants' rights took place during Design and Furniture Week (see the introduction) – had in the previous three months witnessed another protest for immigrants' rights, the display of catwalks to celebrate Fashion Week, a Carnival parade, the demonstration of old street games for children, and a rally by left-wing youth activists – all happening against the backdrop of the kiosks set up by right-wing local political parties, information tables set up by non-governmental organizations, and the huge advertising signs by international clothing brands that take a good portion of the wall space of its buildings.

Because public spaces like this piazza can enable encounters, social debates, and political participation – or at least maintain these as imaginary possibilities – scholars have argued that public spaces can encourage diversity and equality in urban locales. One of the central features of public space is that it ideally provides a place where people can engage with others who are part of the same society, yet different from them, and are thus confronted with an alterity that asks for recognition.[1] Giusti (2008), for example, points to recreational activities in piazzas, parks, and streets in Milan as possible pedagogical moments and as opportunities for encounter between Milanese of different cultural backgrounds and between Italian- and foreign-born residents.

Public spaces are one manifestation of the public sphere, described by Jürgen Habermas as "a realm of our social life in which something approaching public opinion can be formed" (in Habermas, Lennox, and Lennox, 1974: 49), and to which everyone can have access. This public opinion, which Habermas sees emerging from "the critical judgment of a public making use of its reason" (Habermas, 1989: 24), can balance the ruling powers of the state and keep it accountable to its subjects. Although the public sphere cannot be limited to squares and piazzas, as it encompasses many social domains including virtual networks and the media, public space is a crucial part of the public sphere because this is where most easily and directly "private individuals assemble to form a public body" (Habermas et al., 1974: 49).

Here the role of public space as a site for civic engagement – sometimes expressed using the concept of the agora, the idealized Greek market square where issues of importance were debated by the residents of the *polis* – was an important aspect of streets and piazzas to many people in Milan (whether or not they agreed with the movements that used them). One day, for example, I met a group of retired union leaders and members who had organized a campaign to have the streetcars of one of the transit lines, the number 16, replaced by newer bus models without steep steps so that elderly inhabitants could board them more easily. For this campaign, two men were wearing sandwich boards and walking on one of the streets of the centre, and two women were sitting at an information table. As they explained to me during a long conversation, the tiny group was also planning a "rally" to take place a few days later on the transit route of the 16 itself.

It seemed fitting that prior union activists would spend some of their time campaigning in favour of elderly bus users; however, what surprised me about this encounter was their strong sense of entitlement to use streets as a medium for protest and change, irrespective of the fact that the group comprised only a handful of people. Another everyday instance of a piazza becoming a locus for social critique – although with a twist – is the following encounter with a storyteller:

In the Duomo Piazza an older man with a guitar is preparing to sing a song, with a crowd of onlookers cheering him on. As I come closer, I see that he has a large moving billboard as a background, announcing that he is a storyteller, and that his specialty is singing satirical songs about politicians. The titles of his music tapes reveal the main object of his critique: Berlusconi. As he tries his sound system, tunes his guitar, and starts telling the latest jokes about Berlusconi, suddenly a very loud noise from behind startles us all. The storyteller is quick to remark: "It is the Power! It is the Power [of the Institution, of the Status Quo]! See? We are in the piazza, but here immediately the Power ..." his comment trails off, as the crowd laughs, and he starts singing his song. What did the storyteller want to say? Perhaps that no matter what story we tell – especially if we are in a piazza – the power-that-be is always listening in, and always louder than our guitar? Or that, even if Power is always looking over our shoulders, as long as there is a piazza, we can go on singing our songs? (10 Dec. 2004)

I do not want to suggest that public space is invariably used for progressive causes or that its appropriation by ordinary citizens necessarily sustains equality and justice; indeed, often the reverse is true. The extreme right, and outright fascist groups and political parties routinely claim the streets in Milan, sometimes even performing Mussolini's Roman salute during their marches. Indeed, the very reaction of the storyteller at the sudden interruption of his performance was a reminder that precisely the indeterminacy, openness, and accessibility of the piazza is what enables it to lend space to different, and at times opposite, perspectives and actions (Navaro-Yashin, 2002: 1–2).

Still, what is important to note here is that public space can at times serve (for both scholars and city inhabitants) as an imaginary horizon. What emerges most forcefully in much of the literature is that public space can work as a powerful, even necessary ideal to hold on to[2] (Low and Smith, 2006; Caldeira, 2000). In fact, although Habermas has been criticized for his inadequate analysis of gender, overprivileging of rationality, and discounting of performance and theatricality as a tool for progressive political action,[3] according to Calhoun, one of his most interesting contributions was to point to public space as an ideal arising from "flawed material practice" (1992: 39). Understanding how "socio-historical conditions gave rise to ideals they could not fulfil" (40) could encourage people to address this discrepancy and lead to social change.

The notion of public space as an available, although unattainable ideal became useful during my fieldwork when it gave rise to a set of ethnographic questions: Who is motivated by particular understandings and ideals of public space – how, why, in what circumstances, and with what consequences? How do concepts and ideas about public space come alive, and in this process, how do they acquire different meanings? These questions are especially significant because public spaces can become tools of exclusionary practices, in spite – and at times because of – their supposedly liberatory roles. This, in fact, was another aspect of streets and piazzas that became important during my research. Public spaces often become sites for discrimination and control, and thus often they are not conducive to the pedagogical encounters hoped for by Giusti, or the rational debate encouraged by Habermas.

Near Milan's Central Station, for example, a little forest of yellow metal bars prevents homeless people from sleeping on the large vents placed over the underground passages to the subway, which release warm air. And in the Duomo Piazza, a city ordinance forbids people from sitting on the steps of the cathedral. From my experience growing up in Milan and meeting friends in the piazza, however, this rule is enforced differently depending on who sits there. White women are only sometimes asked to move from the steps, but visible minority residents, and Black street vendors, in particular, are very quickly approached by police. This shows that the ways in which people inhabit and use public space affects and is affected by their status and participation in the wider society (Tonnelat, 2008; Low and Smith, 2006; Teelucksingh, 2006; Mitchell, 2003; Razack, 2000; Holston, 1999).

Examples abound in other cities of Italy as well. In Turin's San Salvario neighbourhood, examined by Maritano (2004), a grassroots citizens' committee organized rallies, festivals, and public forums in order to minimize the immigrant presence in their streets, generating wider social and political responses. And Melossi (2003) describes how in Emilia-Romagna the percentage of visible minority immigrants stopped by police randomly on the street is much higher than of white Italians (see also Dal Lago, 1999: 42). Similarly alarming, in 2009 and 2010 many cities in Italy, including Milan, were debating whether to establish citizens' patrols that would monitor the streets at night. Although the alleged goal of these proposals was to provide safety for residents, they were criticized as a way in which private citizens are given power to decide who should be in the city and when. The widespread worry, in particular, was that they would be used to deter migrants from using city spaces.

As these examples attest, the limited entitlement to public space by less desirable residents has to be also understood in the context of the rising concern with security and the neoliberal restructuring of the city (I discuss these issues at greater length in the next chapter). Analysing the approach of the Milanese centre-right administration under Mayor Moratti, Bricocoli and Savoldi (2010: 207ff) argue that public spaces, especially green areas, are increasingly conceived as quantifiable quotas for a group of specific inhabitants. The ideal to render them accessible to all, and significant for civil society, is left aside. Instead, public space takes on the function of a "buffer zone" (216; see also Paone, 2008), an open area that provides deserving inhabitants with a confortable separation from groups perceived as different or dangerous, and from city problems such as traffic and congestion. This is especially the case for parks, which Bricocoli and Savoldi call the "green of distance" (2010: 211). There is little or no effort to integrate the city, to work with its complexities, or to value the flexibility and multipurpose character of spaces and crossroads. On the contrary, the goal is to compartmentalize places and people.

Bricocoli and Savoldi's observations are illuminating because they interrogate the very meaning of public space and its strategic deployment in contemporary Milan, where new urban developments often market themselves as initiatives that will create new, improved public spaces. The public space deployed by the developers is evidently not the same thing as the public space understood by the storyteller, by the retired union organizers, by the priests facilitating recreational activities on church grounds, or by Northern League members who organize debates in the streets to reclaim "lost" neighbourhoods – to name just a few.

This emerged very clearly from my encounters with different people, social groups, and organizations in Milan. Everyone I talked to could easily refer to public space as a meaningful category, an important aspect of city life, and a series of material places (although as varied as clinics, parishes, clubs, and piazzas); most often, however, this was only a point of departure – a beginning of many interrogations. As important as the actual locales they pointed to, and the reasons why they did so, were the various ways in which "public space" brought different topics, positions, and stories into the conversation. To say it differently, critiques and careful celebrations of the workings of public space are very helpful in tracing the different entitlements of inhabitants. However, one of the shortcomings in some of this literature is that it easily crystallizes the very term "public space" as a bounded sociological object that can be used by city inhabitants. During my fieldwork, however, I found that it was often the unravelling of this category that made it into a helpful lens for understanding contemporary issues, concerns, and hopes, as debated by Milan's residents.

For these reasons I wish to argue that more than an object of inquiry, public space is best approached as an ethnographic possibility. It is an entry point or a perspective: a lived and ethnographic location from where certain ideas, struggles, practices, and social realities become intelligible for both the ethnographer and the local inhabitants. This is both a methodological and a theoretical stance: it is not just that I, as a researcher, use public space as a material reality and a concept to understand Milanese local concerns. People in Milan themselves deploy public space as an opening for interventions; see Pigg (1996) for a similar approach regarding shamanism and modernity in Nepal.

Milanese inhabitants do more than talk about, journey through, and/or occupy streets and piazzas. They deploy public spaces as tropes, imaginaries, concepts, and anchors for memories in order to make statements about a variety of other things such as gender relations, politics, or the future. As such, public space acts as a medium for discourses – not only as a physical space to debate issues, but also, and especially, as an idea that makes conversations possible between different people and positions, while at the same time embodying these differences.

When I started my research, for example, I was surprised that one of the very first things that came to my respondents' minds when I asked them about public spaces was the issue of immigration and the place of migrants in Italian society. The idea of public space became for many people I talked to a way to articulate a perceived difference between themselves and others and to discuss multiculturalism and social inequality. In this way, public space became a concept people could refer to in order to enter the discussion. What made it into a particularly important and charged discursive field was exactly that different publics brought forth various ideas about culture, politics, "Milanese-ness," and what public spaces should be about.

To say it differently, and borrowing Pigg's (1996) observations on the meanings of "modernity" in Nepal,[4] what public space is, and why it matters, depends on the social positions of those who inhabit it because its meanings and roles are redefined every time it is used, narrated, and commented upon. For example, people involved with the Milanese Social Centres movement talk about public space as a self-governed locale, gained by squatting. The separatist right-wing party Northern League refers to public space as a property of the "legitimate" citizens: white Italians. Don Felice and Mohamed Ba, on the other hand (see chapters 5 and 8, respectively), highlight the diasporic connections that layer public space to problematize simple encounters between a priori established subjects.

In all of these examples, public space is a circulating idea and object of reference that becomes something different as it is deployed and takes hold. Through this very process, however, public space becomes a dynamic and shared terrain for engagement – it provides, so to speak, an avenue for ideas to travel. What is especially interesting is that public space works as a connection between places, discourses, and issues not because its meanings, and the experiences associated with it, are fixed or stable, but rather precisely because of the many displacements, misunderstandings, and reinterpretations that it engenders; see, for example, Tsing (2005) for a similar approach. In this context, as Pigg writes, "ethnography [...] can be a method for tracing the situated practices through which" locally significant terms like "public space" are "asserted while making evident the displacements such assertions produce" (1996: 164).

Some recent urban anthropological studies encourage a similar line of thinking when discussing how particular spatial arrangements and behaviours in public space can become ways to represent and signify modernity and cosmopolitanism[5] (De Koning, 2009; Harms, 2009; Newcomb, 2006; Schielke, 2008). In Cairo, for example, while conversing with unrelated men is morally dubious in most parts of the city, mixed gender socializing is acceptable for a woman if she is of a higher class and in a respectable place (De Koning, 2009). In turn, the spaces in the city where this can happen – the cafés and neighbourhoods frequented by upper middle-class women who are part of a new professional sector of the population – become modern, cosmopolitan, middle-class spaces. What interests me here especially is that, ethnographically, particular enactments of public space become privileged locations from where to see "modernity" and "cosmopolitanism" and trace them as practices and as imaginaries. In turn, looking at the latter as a set of everyday behaviours, relations, and discourses reveals how public space is a constructed terrain of struggle through which people take part in and contend with locally relevant ideas and understandings.

Significantly, notions of modernity and tradition are at work in Milan, too, where certain ways of using public spaces are often seen as more modern or cosmopolitan (these two categories are often conflated in common discourse in Milan) than others. A young, upper middle-class man I interviewed in 2009, for example, explained that one of the latest trends for people who wanted to be seen "on the cutting edge" is to take cooking courses in a Milanese locale that is in full view of the street because of the floor-to-ceiling windows. This way of being both present in and removed from the street, while engaging in sophisticated food preparation and tasting, markedly contrasts, for example, with picnics in the public parks. The latter, especially if the people participating are of immigrant communities, are seen not only as simpler and/or more economical ways of enjoying the city, but also as more "traditional" or rural uses of public space, thus signalling a lack of modernity – understood here as refinement.

Similarly, many of the middle-class Milanese I talked to described weekend window-shopping by the inhabitants of the city's hinterland as old-fashioned and provincial. Importantly, all of these practices directly engage the politics of seeing and being seen that are such an inherent part of Milanese city life. In this process, specific spaces and distinct ways of being in them become tools for asserting class membership, sophistication, and cosmopolitanism – and distance from those who are perceived as backward. Examining these particular enactments and uses of public space helps us to understand discourses and practices of distinction (Bourdieu, 1984) and other important aspects of an urban culture as they are enacted in everyday life. In turn, attending to these ideas, and seeing how they circulate and are deployed by city residents, helps us appreciate public space as a terrain invested with shifting meanings and social practices, often in conflict with one another.

The Visual Lives of Public Spaces

Practices like taking cooking lessons in full view from the street underline the complex role of seeing, appearing, and being seen in Milan. As a social practice with political consequences, vision is significant both for understanding streets and piazzas as dynamic landscapes and for addressing everyday negotiations and relations between different inhabitants. The following images introduce some of the ways in which visuality emerges in, and affects, Milanese public spaces, and participates in people's relationships with the urban terrain.

Difference and Visibility in the Duomo Piazza

Figure 2. A "Chinese bride." (March 2005).

What makes the bride in Figure 2 "Chinese" in the eyes of many Italian-born Milanese is her very presence in this particular space. Many people in Milan are amazed at the fact that a new bride would go into the Duomo Piazza and "get her dress all dirty" among the pigeons. This practice by some residents of Milan is influenced by Chinese bridal customs of visiting status places in the city (D. Cologna, personal communication, 2005); nevertheless, all complexity, historical variations, and contextual specificities are lost in the simple characterization "Chinese bride."

Figure 3. A "group of immigrants" hanging out and chatting near the entrance of the subway. (March 2005).

Several of my interlocutors pointed out "groups of immigrants" like the one shown in Figure 3 as a very obvious presence in the Duomo Piazza. But why are they so noticeable? Their visibility arguably derives from their use of public space (it is usually younger people and not middle-aged men who stand near subway entrances), as well as their appearance. In this context, the latter does not simply mean what they look like, but rather the way in which they participate in the shared visual landscape. Compare them, for example, with the three men on the right side, who are "looking Milanese" by walking purposefully and well dressed right in the middle of the piazza.

"Counterarguments"

Figure 4. Posters carried in a rally against the Security Package, a proposed series of measures and laws that restrict the rights of immigrants and target illegal immigration. The identical images hold each a different caption: *Italiano*/Italian, on one side, and *Straniero*/Foreigner on the other. Other signs showed identical big splashes of red ink, with the alternating captions "African blood" and "Italian blood." These striking words and images were a bitter comment on the proposed new requirement for doctors and nurses to report illegal immigrants if they came to hospitals to seek treatment. (Photograph by Monica Muraro, Feb. 2009.)

What if we all became immigrants? What if we all became Italians? Can we really tell who is Italian and who is not? What does it mean to look Italian or to look "other"-wise? How does the law shape what we see

and how we can show ourselves to others? Interestingly, these signs
(see Figure 4) promote equality and social inclusion by using a technique similar to the profoundly discriminatory campaign of the right-
wing party, Northern League (*Lega Nord*).

Figure 5. Poster during the 2011 elections. The image seems to ask: If immigration continues, will all Milanese become Muslim? And can Muslims ever
become "true Italians"? (April 2011).

The Northern League posters (see Figure 5) that were present on city
walls in the same period as the rally against the Security Package, as
well as during the 2011 election campaigns, also depicted Italian- and
foreign-born residents becoming alike. In contrast to the signs used in
the rally (see Figure 4), however, they make the point that immigrants
can never become "real Italians," and in this way evoke feelings of fear,
absurdity, and cultural loss.

"Caring for" Public Space

Figure 6. Armani advertisements hanging from the side of a building border-ing a small green public space, in March 2005 (*bottom left picture*; still frame from a video recording at the site) and in April 2014 (*top and bottom right pictures*; photographs by Aurelio Bonadonna). In 2005, on the left side of the grassy area, a telephone booth was also covered in an Armani ad. The small white sign on the grass reads: "Here the green [the grass and the plants] is cared for by Emporio Armani."

In Figure 6 the overwhelming presence of Armani in this piazza, for years, constructs and deconstructs public space. Fashion, by investing (in) and inhabiting public space, influences the city landscape. Conversely, it is often by meddling with the visual feel of a certain place that advertising (in various forms, including screens playing videos from the windows of shops and from the façades of buildings), retailing, and other aspects of the business of style transform and affect public space.

Transparent and Porous Divides

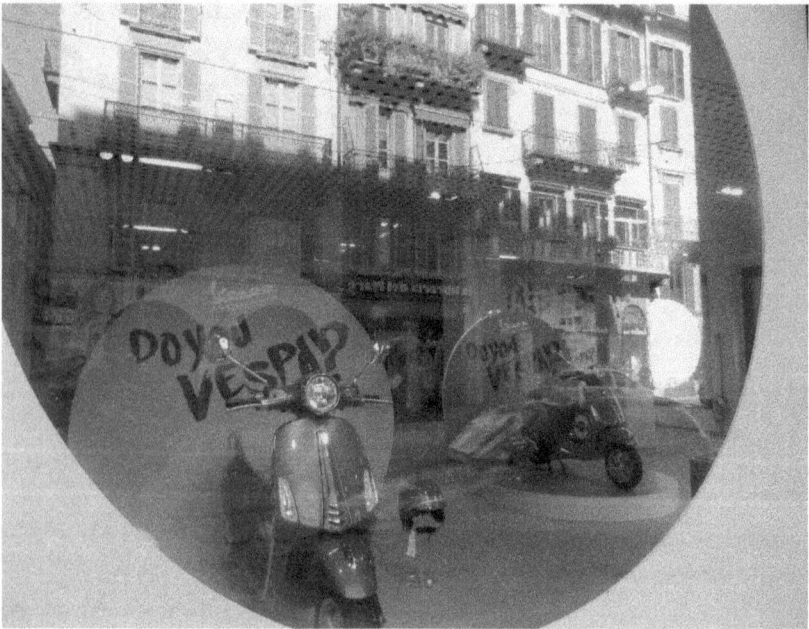

Figure 7. Reflections in a motorcycle store window. (Photograph by Aurelio Bonadonna, April 2014).

Where does public space end, and where does the store start? As Francesca, one of my guides, taught me, shops and streets mirror, "borrow," and embellish each other, playing with the very boundaries between

Figure 8. A clothing store in the central area of Milan. (Jan. 2005).

public and private (see Figures 7 and 10–11). Michele De Lucchi, for example, who designs both service and retail shops in Italy and internationally, describes his role as an architect to "create a rhythm," a "game of opening and closing" between the stores and the streets (see Figure 8). He uses glass panels and windows as a way to create "spaces that seem an extension of the sidewalk and that enter into the architecture, and spaces that, in contrast, establish a complete separation between the material [of the street] and the inside [of the stores]." According to De Lucchi, "the real issue is not closing, but rather opening, and in all my projects there has been a commitment to take away the burden [lit., the embarrassment] of the threshold" (De Lucchi and Villani, 2004: 73). See Figure 9.

Figure 9. Interestingly, one of the latest trends in the city in 2011 was to place little benches in front of stores, inviting people to enjoy the particular private/public space of the threshold, and to become part of the shop's display. (June 2011).

Figures 10 and 11. The interplay between stores and streets interpellates city dwellers at both sides of the glass divides. Here, people inside a boutique appear to onlookers, as passers-by in the Duomo Piazza are reflected by the display window and seem to become part of the interior space of the shop (photograph by Aurelio Bonadonna, April 2014).

Figure 11.   A woman outside a pastry store seems to look out from inside its window. (Jan. 2005).

Figure 12. A composite image of Gucci, in the Galleria Vittorio Emanuele. This Gucci consists of a fashion store and a connected sidewalk café, where waiters wear black and look/move like models and/or actors. (Jan. 2005).

The Gucci café makes the shop spill into public space (in fact, claiming and fencing a part of the most central of all promenading spaces in Milan!) and creates an opening from the street to the interior of the store – almost inviting public space in (see Figure 12). Wilson (1987) associates (window) shopping with theatre and spectacle. An interesting twist here, however, is the uncertain location of stage(s) and audience(s). Are the people drinking coffee in full view of the promenading crowd (and indeed in the same public space), attended by models/waiters/actors part of the show? And what about the promenading crowd itself? Who (and where) is the audience, and who (and where) are the performers? And who does not get to (or refuses to) participate?

Dressing the City

Figure 13. Rally in the centre. (Jan. 2005).

In today's Milan the representations of elegance and allure are so ubiquitous and so significant visually that one has the impression that the fashion culture is always already part of the urban scene (see Figures 13 and 14). Whether you are having coffee with friends, going to church, making a phone call, or participating in a demonstration, you are always doing so under the gaze of Armani (or another major fashion brand).

Figure 14. Advertisement on a church. (June 2011).

Figure 15. Milan's Central Station. (April 2009).

Milan's old Central Station (see Figure 15) has been redesigned to look like a shopping mall for affluent travellers. Yet, the Central Station has always been an important public space for new immigrants: it has been a way to connect with people, to reach the city, and it is a workspace for pedlars. Is this shining "restyling" (Burigana, 2009: 25) a way to displace one kind of traveller in favour of another? Is the sense of seeing associated with window-shopping and the glass panels a way to try making invisible or out of place low-income users?

The remodelling of one of the subway stations in the centre of Milan, and close to one of the major redevelopment zones, is promoted as a way to render it more in tune with its location in a fashionable district and in a fashionable city. The image in the poster (Figure 16) shows the silhouettes of people transitting through the station in a striking resemblance to models on the catwalk, and the description reads that the changes were "inspired by the city's catwalks." But who will be fashionable enough to take part in the new spaces being constructed?

Figure 16. A poster by the municipal transit authority. (April 2009).

Invisible People in Invisible Places

Figure 17. "Empty" areas in the north of Milan. (Feb. 2009).

The dismissed areas (*aree dismesse*) can be seen as existing at the cross-roads between different (in)visibilities. Although massive and numerous, they often remain invisible from public view and also render invisible those who inhabit them (in the case of the picture that is Figure 17, a group of Roma people), and those who hope to profit from their redevelopment.

Figure 18. Urban redevelopments in Bicocca. A new building looks onto vast spaces of nothingness, as if waiting for something to happen, or trying to divine an uncertain future. (April 2009).

The images in Figures 2 to 18 suggest that in contemporary Milan visual elements and practices matter in several different yet interlocking ways. These include alternating gestures of recognition and othering, the uncanny presence of spaces that conceal social relations and their subjects, fashion and aesthetics as a material and imaginative force, and an everyday, shared habitus of seeing and being seen. In discussing these different aspects, what interests me especially in this book is how these multiple and intersecting ways of seeing, appearing, and concealing mediate people's participation in the city and thus help constitute particular spaces and particular inhabitants.

Robert Rotenberg's study of European cities at the turn of the twentieth century offers a valuable point of departure for my inquiry: he describes how in Vienna, Paris, and London the rising middle classes sought to reframe their city "in a bourgeois image" (2001: 7) by creating a particular visual landscape, a representational and symbolic space modelled after the Paris arcades. In this process, the most visible barriers in the city became the most transparent ones: the aesthetics of desire that separated classes permitted some people to be shown and others to look on:

> The Paris Arcade (1822–37), the first "mall," used glass windows to display luxury goods, especially fashion, in ways that evoke fantasy and desire [...] The transparency of the windows separated the real from the imaginary and the consumer from the object of desire. The approach was soon imitated in other cities. Owning such a suit of clothes converted the fantasy to concrete signs of class membership. To participate with others in the ownership of such fantasies was to realize an identity [of being upper middle class] that was truly exceptional and enormously satisfying [...] In the Arcade, the glass window accomplished the boundary between the object and the one who desires it. In the city, other mechanisms would have to develop to create symbolic transparencies, enabling some to consume and others to merely watch and desire. (Rotenberg, 2001: 11)

In cities such as Vienna and London, this was done by creating new, clean, respectable neighbourhoods and separating them from poorer areas of the city that were cast at the periphery. "In the end," what was "made possible was the consumption by some urban residents of street addresses and other commodifications of space that fed their sense of exceptional identity, while others, both co-residents and provincials, could merely look and desire" (ibid.).

Although Rotenberg analyses European cities in the past, his observations prove illuminating for contemporary Milan, where the new neighbourhoods being constructed market themselves as prestigious signs of membership in the city of fashion, and where the urban landscape itself has become a medium of attunement to global sensibilities and ideas of the future. Common discourses in Milan on its revitalization projects have described them as an "urban renaissance" (see Balducci, 2007: 7), as crucial steps for building the city of tomorrow, and as a response to the inhabitants' "need and pride to see Milan [...] in step with contemporaneity," that is, with modernity (Boschetti, 2010: 26). In this context, the growing skyline of Milan takes on a particular role, representing the creative and dynamic city, one that seeks to attract affluent residents and professionals. Here, as in Rotenberg's analysis, the landscape of the city, rather than an inert backdrop, is more fruitfully addressed as a process (Weszkalnys, 2010), in which inhabitants participate in differing ways. To use Gillian Rose's words, it is a "way of seeing which we learn" (1993: 87) and which helps construct particular relations between people and places (see also Blomley, 1998).

What I find particularly interesting in Rotenberg's work is his insight that those transparent divides are distributed within the city. It is the very lives of people, their positions, and movements that draw together vision, aesthetics, and desire so as to (re)construct, shift, and reinterpret boundaries within the urban terrain. During my research I was intrigued by how my walking guides used practices of seeing to imagine and situate themselves in Milan's social landscape and to participate in its public spaces. They pointed to landmarks to show their connections to particular locales, and they invited me to see absences and voids as signs of the past and/or the future, and commented on people's appearances as indications of their social positions. In this way, seeing and situating oneself into the cityscape became a way of relating to other inhabitants.

Let me emphasize here that by saying that vision is an important aspect in people's relationship with public space I do not intend to marginalize the other senses involved in urban life. On the contrary, I am interested in vision as a practice that is dynamic and embodied, connected to other sensual experiences of the city, and shaped by the very movements of the seer within the landscape, rather than a distant and distancing view. It is a vision that makes the body in space matter. I argue that such a framework is especially relevant in Milan, considering the ways in which seeing and appearing operate in public space – including the legacy of the promenade, the important effects of fashion and design as a cultural, economic, and social force, and the connection between alterity and (in)visibility.

Christopher Pinney (2004; 2002), writing about visual cultures in India, similarly argues against vision as being necessarily a distant, static manner to relate to the world, and he urges anthropologists to pay attention to its "mutuality and tactility" (2002: 355). He calls the first kind of vision, which is "disembodied, unidirectional, and disinterested" (359), "anaesthetics" (2004: 19). On the contrary, "corpothetics" (ibid.) are practices of seeing that connect the looking body to the one of another person (or deity, as in the case he discusses). Pinney talks specifically about *darshan*, the religious viewing of the images of gods and goddesses; however, the "corporeal visuality" (2002: 356) he describes is very useful for understanding Milanese practices of looking that, instead of distancing the body, link it to the presence of others and to their journeys in public space.

A notable expression of this is the old practice of the promenade, or *struscio*[6] (lit., the brushing of one's body against others): the leisurely strolling in the central streets to see people and show one's (good) appearance, found in many Italian cities. The historian and novelist Carlo Castellaneta describes the traditional struscio in Milan like this: "in the habit of the promenade there was a double pleasure: the one of looking and the one of being looked at, of greeting and of being greeted. The street became a catwalk where one could show off a new dress or a tan" (1997: 99). Although at first sight this might appear an exceedingly mundane occurrence, we should not take the struscio lightly. Giovanna Del Negro, for example, argues that the *passeggiata* (another word for the promenade) in her fieldwork site, a central Italian middle-sized town, provides an embodied and dynamic snapshot of the town's concerns, its history, and its social divisions. Because it is a shared performative practice where residents "explore the meanings of gender, class, age, and generation" (2004: 39), it not only reflects unequal power relations and categories, but also provides an arena for city dwellers to question them. In Genoa, the promenade discussed by Emanuela Guano (2007) has a different name – *fare le vasche*, meaning literally swimming up and down the lanes – and different dynamics, but it is nonetheless a crucial way in which social identities are created, maintained, and commented upon.

Milan is no exception. Tied with concepts of gender, fashion, class, and diversity, the struscio entails an active seeing of people in urban space and the performing of one's self so that it can be seen and become part of the social terrain. Although the traditional struscio is usually linked to particular places and times (such as the central streets or piazzas of the city, on the weekend and in the evenings), and is not as popular as it used to be forty or fifty years ago, it still frames many people's promenading practices in the central avenues and piazzas of Milan. Moreover, as a wider visual culture, it permeates life in the streets. Here is Castellaneta again: "Our [Milanese] women have [...] a completely feminine ambition to be admired, for their figure, for their dress, and for their comportment [...] And this not only can be noticed in the streets of the centre, where one promenades as if [on display] in a store window, but also in the subway where it is rare to meet a girl without makeup or a woman clerk dressed carelessly. And so many times I noticed a housewife shopping in a supermarket dressed in a [very expensive] fur coat!" (1997: 35). As indicated by Castellaneta's words, the struscio has often worked to strengthen traditional ideas of femininity, as women are usually expected to conform to hegemonic canons of beauty, heterosexual attractiveness, and femininity. Needless to say, women at times follow the above ideals, at times play with them, and at times actively resist them.[7]

In contemporary Milan not only clothing and bodies, but also cars, cell phones, and accessories are important parts of the show. According to one of my guides (a young working-class woman), for example, "shoes are essential" for daily life in Milan. And another young, middle-class woman recounts:

The [plastic or paper] bag [people carry with them] corresponds to their clothing, to the type of person [...] Because there is a difference between bags. If you carry around a bag from Prada [a very expensive fashion store], black with the small logo of Prada on it, is very different than if you carry around [...] a bag from Oviesse [a popular supermarket]. Even the bag has its role in the choreography of dressing. (12 March 2005)

Although not part of the promenade per se, many other "fashionable" behaviours and objects are intimately linked with these techniques of being and appearing in front of others and within city spaces. In 2009 and 2011, for example, I found that food had emerged as a key "accessory" in city life. It had become fashionable to take refreshments and snacks on spoons and from glasses rather than plates – as if they were drinks – and to consume salt of different colours. And going for happy hour in particularly popular locales had become a new "ritual" for many Milanese. Indeed, when I interviewed a young, affluent, fashion-conscious man in 2009, it took him over an hour just to list what had become trendy in the city at that time!

Ways of walking, being, and seeing in public space, such as the struscio with its constant attention to others – became interesting for my research for several reasons. First, these practices were available to inhabitants both as topics of discussion and as performative forms for taking part in public spaces. This double role, in fact, made them into particularly interesting conjunctures for inquiries into urban life. Consider this example: several fashion-abiding, white, middle-class Milanese women I talked to insisted that they do not engage in the struscio because "they are too busy," because it is too "provincial" a practice, and because Milan has always been "an introverted city" (and thus "real" Milanese do not like to show off their attire). All of these women, however, dress up very carefully and enjoy strolling in the streets to see people and to participate in the urban scene.

Similarly, other white, Italian-born Milanese women I talked to said that they *never* go to the centre, even if they actually do. Several of them explained that they do not go there and they do not engage in the struscio because, especially on weekends, the centre is full of *zarri* (people from the hinterland who can be recognized as such because of their way of dressing) and *extracomunitari* (migrants arriving from outside the European Union). Rather than indicating that the struscio has disappeared, those comments point out that, for some women, the struscio might no longer be a "Milanese" thing to do – or, conversely, that *not* engaging in it might be a "Milanese" thing to do. In other words, comments such as the ones above may express an imaginary that only "Milanese" people (in contrast to immigrants or inhabitants of the periphery) are truly interested in beauty and fashion yet cosmopolitan enough to *not* participate in the struscio. At the same time, expressing a distancing from the struscio did not prevent my interlocutors from actively participating in its visual culture.

Second, as this example indicates, the seeing of the struscio – in which one's body is connected to others through an intricate maze of looks, reflections, and social judgments – provides a wider framework for expressing and negotiating belonging and identity in Milan. Here we should remember that the struscio can, at times, work as a particularly apt venue for discourses on multiculturalism and identities precisely because it is a relational, dynamic, reflexive, and performative practice of seeing. It is a form of looking that requires another gaze. Because the struscio links people with a particular context and crowd, it becomes very important with whom and where it happens. It is interesting in this context to note that Castellaneta (1997) traces the beginning of the relative decline of the struscio in Milan to the entrance of lower-class people to the centre of Milan, originally the space of the elite and the upper classes.

In his famous description of the lives of urban places and people, Michel de Certeau criticizes the power of the planner over the city as a disembodied view from above:

> The desire to see the city preceded the means of satisfying it. Medieval or Renaissance painters represented the city as seen in a perspective that no eye had yet enjoyed. This fiction already made the medieval spectator into a celestial eye. It created gods [...] The totalizing eye imagined by the painters of earlier times lives on in our achievements. The same scopic drive haunts users of architectural productions by materializing today the utopia that yesterday was only painted [...] The panorama-city is a "theoretical" (i.e., visual) simulacrum, in short, a picture, whose condition of possibility is an oblivion and a misunderstanding of practices [...] The ordinary practitioners of the city live "down below," below the thresholds at which visibility begins [...] [T]hey are walkers, [...] whose bodies follow the thicks and thins of an urban "text" they write without being able to read it. These practitioners make use of spaces that cannot be seen; their knowledge of them is as blind as that of lovers in each other's arms. The paths that correspond in this intertwining, unrecognized poems in which each body is an element signed by many others, elude legibility. It is as though the practices organizing a bustling city were characterized by their blindness [...] Escaping the imaginary totalizations produced by the eye, the everyday has a certain strangeness that does not surface, or whose surface is only its upper limit, outlining itself against the visible. (1984: 92–3)

This passage from de Certeau, as well as the work of many other scholars, criticizes vision as a distancing act of power. Donna Haraway, for example, calls it the "god-trick" (1991: 189) of seeing everything from nowhere, thus emphasizing that the disembodiment of the master subject can often be a way in which powerful interests make themselves invisible. In relation to the city, such a vision can correspond to an urban planning imposed from the municipality or from developers and that the "practitioners of the city down below" cannot easily grasp or influence.

Although I find the analysis of hegemonic vision as distancing and disembodied very useful, I find de Certeau's description a little disconcerting. For de Certeau defines the people who actually live in the city as blind – involved in activities that "elude legibility" and visuality and that are more akin to speaking than seeing. The lack of vision is used here, as a metaphor, to talk about the resistance of people and of experiential daily life against hegemonic forms of control and governance. However, why do the walkers necessarily have to be blind? Why can they not be engaged in all kinds of different visions and acts of seeing – some of them resistant, and some of them not? And perhaps all of them related in interesting ways? Disembodied, distancing vision could be just one type of vision in city spaces, which perhaps could at times be used both by the planners in charge of "space" and by the walkers who create "places."[8] The practices of looking/walking employed by my interlocutors and that I discuss in this book, as well as the complex politics of seeing, not seeing, being seen, and not being seen described by several organizations I met with in Milan require us to think of multiplicities of gazes and of the relationships between vision and embodiments – not just disembodiment – in city space.

As a strategy to examine these processes, I suggest that we pay attention to what we could call, borrowing Bakhtin's word (1981: 263), the heteroglossia of vision. By this I mean the way in which images and modalities of looking and appearing in the city inhabit the same locales – being in a sense superimposed on one another and existing together with many other ways of seeing, as well as with performative and narrative articulations of vision and aesthetics. Consider this example: in 2007, in Corso Buenos Aires, a major shopping avenue in Milan, framed on both sides by international brand name stores such as Benetton, Stefanel, Zara, and HLM, an electronic goods retailer hired two models to squeegee for a day in order to advertise its products. The two blonde women, dressed in tight shorts and tank tops sporting the Kenwood logo, stood at a set of lights and washed the windshields of cars that were transiting through the Corso (*La Repubblica*, 2007).

Although the women's work and the looks they elicited belonged to the universe of the catwalk, and actresses in commercials,[9] it is hard to ignore the way the models reminded passers-by of the regular, familiar squeegeers in Milan, and of the very different reasons and effects that their visibility entails. The latter are usually visible minority immigrants and/or Roma and Sinti. It is interesting that the actions of the two models have supposedly nothing to do with regular squeegeers, yet in fact work precisely because of people's experience with the latter. Indeed, the more we think about it, the more complex it becomes. The pun of the models echoes and changes the asymmetrical looks involved in squeegeeing – something which, from the point of view of the driver, is usually uncomfortable also because of the mutual gaze it seeks to establish and because of the difficulty of avoiding the attention of an "other" who most people wish to be invisible; an "other" who, of all things, insists on clearing the windshield so we can see better what is going on around us!

This anecdote suggests that vision matters in urban locales never in a straightforward way, but rather through complex processes of mirroring, representation, and concealment. Public space, and people's bodies in it, becomes a "dynamic site where many gazes or viewpoints intersect" (Lutz and Collins, 1993: 187). Vision always occurs in, and runs into, complex issues of translations between contexts that are not just connected but often are hierarchically related to one another (see also McLagan, 2002). This means that to understand how vision works in specific contexts, we need to do two things at once. First, we need to reflect on particular visual cultures, situated practices of representation, and the specific social positions engendering and engendered by them. Second, we need to pay attention to how ways of seeing travel and matter across different but connected social spheres.

To return to our example, the visual politics involved in squeegeeing in Milan interpellate people's various experiences with difference, gender, and sexuality. As they travel from one domain to the other, they do different things and have different meanings. Yet, at the same time, the way in which privileged subjects (in this case men in cars, because of the implicit heterosexism of this form of advertising) can deploy vision (can gaze at) consumable female subjects, also confirms that poor immigrant squeegeers are undesirable bodies in public space.[10] Public space here is the terrain that makes the shifting and travelling of vision possible. By taking these dynamics into account, visual practices can be a starting point for analysing how images, (in)visibilities, and lived landscapes help consolidate places and identities.

Chapter Two

Milan[1]

This chapter will acquaint the reader with some of the aspects and dynamics of Milan. Another one of my goals, however, is to foreground the very difficulty of representing urban realities. During my research, people repeatedly reminded me that a city lives in the shadows, too, in people's imaginations, and in the subtle moments and circulations that animate it. In this chapter, therefore, I try to complicate a unified description of Milan, by interweaving my narration with anecdotes, excerpts from my fieldnotes, and parts of a virtual walking tour in which I invite the reader to follow me on an imagined itinerary.

The first one of these "interruptions" is a recollection from my childhood. Regarding the virtual tour, the encounters and events I describe – all actual occurrences from my fieldwork in the city – did not all happen on the same day (they took place from January to April 2009[2]), but they easily could have. I assembled them in one walking journey (giving them a fictionalized sequence) to offer the reader a sense of everyday life in Milan, replete with sudden encounters, changes of trajectories, and emerging interrogations.

While a fictionalized tour is not common in ethnographies (but see Guano, 2003), this chapter takes inspiration from several authors who have been grappling with the difficulty of representing what might not be fixed, completely intelligible, or unified (see, e.g., Weszkalnys, 2010; Doninelli, 2010; Tsing, 2005). As they suggest, there is something to be gained when we pay attention to momentary encounters and remarks, passing affects and movements that do not coalesce in discrete social forms or lend themselves to explanation. These are an "opening onto something" and show "a thicket of connections between vague yet forceful and affecting elements" (Stewart, 2008: 72). Misunderstandings, divergent and multiple translations, and incompatible knowledges are as much and as powerful a part of the social forces as what can be clearly defined or demarcated. For these reasons, the authors from the collective Multiplicity.lab (2007) privilege the collage – in the form of photoessays, annotated maps, and newspaper clippings – as a way to describe the complexities of Milan, and the intersections between its inhabitants.

Echoing the questions and insights of these scholars, the walk I invite you to participate in wants to be a practice of attention to the nooks of everyday life in the city, which manifest important social forces at work in Milan and always exceed scholarly interpretations. Remaining part of the dynamic immediacy of the streets, these instances and encounters spark questions, surprises, and fleeting memories, which do not add up to a neat anatomy of the city but rather produce a "multitude of small shudders in the urban body" (Boeri, 2007).

A recollection: as a young girl growing up in Milan, I was fascinated by stories about the things that disappeared. A few blocks from my high school, in what is now an ordinary block of grey buildings with a good ice cream store at the bottom, some of my relatives could still see – as if it were still there – a tailor shop that used to be my grandmother's, and the railway line running high on a viaduct between the apartment buildings. From one of them, my mother used to watch people in the trains coming by. These tales were both fascinating because they were strange (how peculiar, to have running trains just beside an apartment window!) and deeply disappointing, because they showed me that a part of my city had vanished without leaving me with enough traces to follow it. Twenty years later, a similar feeling of loss over familiar places seems to motivate some of the conflicts over neighbourhoods in Milan. Listening to my relatives' stories as a child, I had always assumed that remembering was an innocent and unambiguous gesture towards the past. Now, I learned that one's feeling of loss might easily lead to claim spaces at the expense of others.

Milan, the capital of the Lombardy Region, is the largest centre in northern Italy and a key "global city" in the European context (Brenner, 2004). Its prominence in Italy, however, is far from just economic. Milan is considered the most cosmopolitan city in the country, and a "symbol of movement, transformation, and modernity" (Bonomi, 2008: 1). This reputation was earned by spearheading some of the most significant Italian cultural, economic, and political changes of the twentieth century, including the "miracle," a rapid phase of industrialization that took place starting in the 1950s.[3] During this time, immigrants from the South as well as from nearby regions settled in the new peripheries of Milan, where the factory became a fulcrum of everyday life. This work was hard and poorly paid, and southerners were often discriminated against; still, industrial employment allowed many families a steady source of income and hopes for the future. The factory, moreover, helped workers from different parts of the region and of Italy to forge alliances, and it became a cradle for activism and reforms.

The "miracle" was followed by a very important period for the history of the city, and one that keeps re-emerging in its collective memory: the time of intense social struggles that culminated in the 1970s, and the so-called Years of Lead (*Anni di Piombo*, or "Bullet Years"). Starting in 1967, university students and then factory workers engaged in concerted actions against the dominant institutions. Many left-wing associations and oppositional groups were started, some of which occupied buildings to protest inequality in the city. In response to the protests and unrest, a bomb was planted in the very central Piazza Fontana in December 1969. The ensuing massacre initiated what has been called the "strategy of tension," a series of violent acts presumably by right-wing forces, but designed to set the blame onto anarchist and communist groups; see Ginsborg (2003: 333ff) for a more detailed history of these years.

Starting in the 1970s Milan deindustrialized and switched to a tertiary economy. In this process design and fashion became two of the most important industries in town. The success of the fashion sector resulted from the combination of several factors: well-consolidated, artisan textile production in the northern area of Milan, a long-standing cultural attention to aesthetics, the presence of independent designers in the region, and Milan's strategic position in relation to European and Italian markets (Crane and Bovone, 2006; Yanagisako, 2002).

This change brought important shifts in the urban fabric. For one, the most prestigious fashion names gradually took over the historic centre, displacing older inhabitants and businesses. For another, the closing of many industries of the "miracle" left huge abandoned areas in the city and its hinterland, the so-called *aree dismesse* (dismissed areas). Pierpaolo Mudu estimates that "by the late 1990s, industrial property across a total area of 7 million square miles had been vacated in Milan alone, not to speak of peripheral municipalities such as Sesto San Giovanni, where closures affected a total of over 3 million square miles" (2004:

921). Some of these empty factories or vacant fields have been appropriated by organizations and groups ranging from political networks to artists' co-operatives. Others have been slowly transformed into the grounds of new neighbourhoods, residences, or facilities for the service sector. As symbols of a previous era, and as valuable sections of the city that can be developed in new ways, they now embody both the past and the future in a complex and conflictual way; see Weszkalnys (2010) for a parallel with Berlin.

Follow me as I walk through the extensive Duomo Piazza, visit the wide promenading routes of Corso Vittorio Emanuele and the Galleria, and weave my way by the very central San Babila and Scala Piazzas. The Duomo Piazza, symbolic centre of the city, is reserved for pedestrian traffic and is a nodal point for two subway lines. It is here that I learned from Mohamed Ba the history of the city. In fact, most of my guides selected this piazza as an important part of their itinerary, and I met many of my interlocutors here, by either approaching them when they were resting on benches, or by being addressed myself. I traverse the piazza's vaste grey expanse – shared by pigeons, people, and statues, and framed by imposing palaces and the century-old Rinascente department store – and walk through the covered passageway Galleria Vittorio Emanuele that links the Duomo with the Scala Piazza. The Galleria is an old, well-established promenading route and is flanked by expensive shops and cafés, although a very heterogeneous crowd of people moves through it. Today there is an additional reason as to why some of these people have come here: there will be a dance to commemorate the centenary of Futurism.[4]

I enter the McDonald's restaurant that, incongruous between high-end boutiques, offers me windows onto the Octagon, the central crossroads of the Galleria, where the dance will take place. The McDonald's is full of a heterogeneous clientele. Middle-class ladies with conservatively coloured, respectable dress, try to eat pastries with their grandchildren, a man with a tie is conducting a business meeting, showing another person some samples in little tubes, and a group of Filipino-Italians are chatting in a language I assume to be Tagalog. When the dance starts, everyone rushes outside to see ten young dancers (men and women, black and white) put on their performance.

They dance in the middle of the promenade, between the onlookers, without a stage or an introduction, carrying music and loudspeakers in their hands. Suddenly, they run away from the scene. Their departure leaves everyone startled. The audience does not leave, but lingers in the space. Although it looks like a regular crowd, there is now something palpably different about it: a feeling of waiting for something that might happen and could transform the place and time we dwell in. The feeling is so strong that it binds us all in a temporary community. In the Octagon, grandparents, babysitters, tourists, panhandlers, passers-by, and businessmen are standing on the four corners, carefully looking at all the others around them. Where have the dancers gone? Will they come back? Who were they and why did they leave? More questions come to mind: Have we always waited for something or someone? Who is coming and who is going? Is the history of the city a sticky spider web of anticipation and nostalgia, a waiting for events that vanish before we can fully comprehend them, and that we forget before ever fully remembering them?

An interesting characteristic of Milan is that its population – about 1.4 million people according to Manuel Aalbers (2007) – approximately doubles during the day because of the very large number of commuters and "city users" of various kinds (Martinotti, 2003 and 1993).[5] For one thing, the great number of commuters is indicative of the city's productive role in northern Italy. Together with fashion and design, Milan is an important centre for finance, publishing, and advertising.[6] For another, the disparity between day and night inhabitants is a witness to the fact that the very high cost of housing has been displacing a large part of the Milanese population outside of its municipal borders.

David Benassi shows that the average rental price in Milan is roughly a third higher than the combined average of the eleven major cities in Italy (2005: 24). The situation, according to Aldo Bonomi (2008) has not been improving, and housing is becoming a heavy burden even for more affluent sectors of the population. The lack of affordable housing[7] has been exacerbated by declining social housing programs and the gentrification of many areas, such as the ones bordering on the Navigli canals, the Isola neighbourhood (see chapter 6), and other working-class zones close to the city centre (Bricocoli and Savoldi, 2010; Zajczyk, 2005). The refurbished and restored Case di Ringhiera present in these areas are a paradigmatic example of processes of gentrification. Previously working-class accommodations, with very small apartments and shared bathroom facilities, many of these buildings gained the (albeit informal) status of historic landmarks. Appealing to ideas of Milanese traditions, and romantic notions of social life as it was at the beginning of the twentieth century, they have now been restored to make room for larger apartments and fatter pay cheques.

Among the people who have the hardest time finding affordable and appropriate housing are immigrants. This not only is because as a group they have less income than Italian-born residents, but also because of discrimination from house owners. Indeed, the Caritas[8] report on immigration estimates that more than 60% of foreign residents (people who are living in Italy but who are citizens of another country) in metropolitan Milan live in overcrowded, precarious, or temporary accommodations (Caritas/Migrantes, 2005; see also Paone, 2008; Bonomi, 2008; and Naga, 2003). Although many immigrants do obtain social housing, because this resource is very limited in the city, it does not significantly improve the situation.

Leaving the Octagon and its waiting crowd, and walking the full length of the Galleria, I reach the Scala Piazza, where the world-famous theatre is located. I walk in via Manzoni, lined with elegant shops and cafés, pass the new Armani megastore with its exclusive hotel that is still under construction, and then enter the gates of the Giardini Pubblici, one of the largest parks in central Milan. The Giardini shows a striking combination of timeless continuity and obvious signs of the recent social transformations. Its café, its pond, the man selling balloons, the children's train and its conductor were already old when I was a little girl. Yet today in the Giardini you can hear many languages being spoken, and nannies originally from the Philippines and from North Africa play with children at the playground.

Traversing, and then leaving, the park I reach Corso Buenos Aires, a wide shopping avenue lined with large stores. As I walk along, I am surprised by how many street vendors are peddling their wares. Most of them visible minority residents, they are selling CDs, umbrellas, bags, belts, and t-shirts from large blankets they have spread on the pavement. The sidewalk stalls become more numerous the further one proceeds from the centre, until the street starts to resemble an open market rather than a succession of Benetton, Zara, H&M, and Mango as in its first section. I arrive in the tiny Loreto Piazza and go underground to take the subway and continue my journey to the north of Milan.

In the subway, as I take a seat close to the exit doors, a man with a violin gets on the train. His violin is repaired with tape, his backpack has holes cut into it to allow the loudspeaker inside to sound through, the bow is already in his hand, and the violin almost on his shoulder. On the other side of the wagon, but facing a different direction, is a policeman. He looks straight ahead, and does not see the man with a violin just a few metres away. The violinist does not look at the policeman at all, but lowers his bow, looks at his shoes, and slowly sits down in an empty seat nearby. The power of the authority – more specifically of the new stricter by-laws introduced in the past couple of years – becomes suddenly apparent in these two looks that do not cross. The violinist is the one who is forced to see – although pretending not to – even when and exactly because he is invisible, or better, one of those inhabitants desired to be invisible by others.

Milan is one of the Italian cities where international immigration has had the most concentration and effects. Its region, Lombardy, is home to almost a quarter of all foreign residents in the country (persons living in Italy but having a non-Italian citizenship). In the municipality of Milan itself, at the end of 2007, the immigrant population accounted for approximately 14% of all inhabitants, while the proportion for all of Italy was less than 7% (Caritas/Migrantes, 2008). Foreign residents registered in the municipality of Milan come primarily from Asia (35.3%), followed by Africa (23.1%), the Americas (21.1%), and the rest of Europe (20.4%). In terms of nationality, the largest immigrant communities in Milan are the Filipino, Egyptian, Chinese, Peruvian, and Ecuadorian (Caritas/Migrantes, 2008: 326). Unfortunately – and notwithstanding the economic, cultural, and social richness they bring – racism and discrimination are widespread in Milan and in Italy as a whole (Merrill, 2011; Dal Lago, 1999; Cole and Booth, 2007).

In Milan one of the factors that might have strengthened a negative perception of newcomers is that foreign immigration to the city has happened roughly at the same time as the difficult process of tertialization[9] (Petrillo, 2004; Melossi, 2003). Most of the migrants to the city who arrived from other parts of Italy in the 1950s, 1960s, and 1970s found work in the growing manufacturing and construction sectors, and many joined unions, but newcomers to the city in the past two decades have found a very different market and employment structure. As unionism reached its lowest level in Milanese history, and the factory ceased to be a crucial point of reference in everyday life, the wide political ideals of solidarity and inclusion promoted by the workers' struggles fragmented into less-cohesive movements and groups that are, consequently, less able to advocate for social equality (Petrillo, 2004). At the same time, the uncertainties brought by deindustrialization have amplified many residents' anxieties towards the changing social composition and increasing diversity of their city.

With the closing of the factories in the period between the 1970s and the 1990s, and the shift towards a service economy, job security has decreased. Although Milan has a very low unemployment rate, this can easily hide the fact that many workers face difficult, temporary, and marginalized labour conditions. Bonomi speaks in this respect of a growing "neo-proletariat of services" (2008: 9), workers who are involved in low-paying, unskilled jobs in the tertiary sector. And Noelle Molé (2010) describes the precarious conditions faced by an increasing number of people in Italy who are casual, or "atypical" workers. In Milan this includes many young persons employed part-time, many immigrants in the service sector, and many workers in the fashion industry.

Consequently, many residents of Milan are not so much excluded *from*, but rather "*on* the labour market" (Andreotti, 2006: 332; see also Dines, 2002). Less-privileged groups such as single mothers and immigrants consistently occupy the lower, and precarious end of the job force,[10] and their being employed simply masks their progressive impoverishment (Andreotti, 2006: 342). In this context, it is telling that although Milan has the highest per capita income in Italy, the poverty level is close to the national average, thus showing that income inequality in Milan is more marked than in other Italian cities. Welfare structures and social assistance, moreover, are scarce and temporary.

The growth of flexible and casual employment and deepening urban inequalities have been part of a more general neoliberal shift that took hold in Lombardy and in all of Italy starting in the 1980s. This included a diminished role for the welfare state and the privatization of state companies, as well as deregulation and trade liberalization. By the 1990s, Italy had implemented (albeit with difficulty and only partially) pension and health reforms, a reorganization of many social programs such as unemployment insurance and disability allowances, a restructuring of some of its fiscal systems, and a deregulation of labour laws (Saitta, 2011; Gualmini and Ferrera, 2004; Ginsborg, 2001). The latter, introduced in 1984, allowed temporary, part-time, and more flexible work to become an important part of the Italian contemporary job market. Molé (2012) describes in this respect a two-tier system, in which long-term employment with benefits coexists with flexible labour arrangements. Moreover, the state shifted responsibility for many public services to the voluntary sector (Muehlebach, 2012).

Neoliberalization in Italy was especially promoted by the political leadership of Berlusconi, who repeatedly proclaimed the need to run Italy like a business,[11] but also it mirrored a larger transformation at the level of the European Union. As part of the creation of a single currency, the EU instituted goals and criteria that curtailed and restricted member states' national programs in the realms of welfare, labour, and unemployment (Hansen, 2000). Inclusion in the EU hinged, too, on structural and social reforms and on fiscal discipline. On a more general level, notes Peo Hansen, the EU relinquished the association between citizenship, inclusion, and the right to employment towards "an increasingly individualized and market-oriented perception of citizenship" (2000: 145).

Climbing the steps from the Precotto subway, a dozen stops from Loreto, and walking along a narrow road, the temptation to take pictures of the "old" meeting the "new" is overwhelming. Just on my right, there is a courtyard with a garden, and just behind it, a house. It is two storeys high and has a long, narrow balcony on its front which connects like an open-air hallway its different apartments or sections. The building resembles the traditional Case di Ringhiera of the early 1900s, celebrated in much of the literature about old Milan. Painted all in white, with a brown tile roof, and with what looks like a fruit tree in its quiet garden, the house also evokes the countryside several kilometres away. Immediately in the background, however, one can glimpse the glittering glass and steel façades of one of several high towers. These are part of the new Bicocca neighbourhood which has been emerging in the last few years, and that includes residences, a theatre, and a large university.

I leave the country house with its garden in front and its towers at its back, and take a streetcar through a tunnel to University of Milan-Bicocca. From the windows of the anthropology department, in a building simply called U9, I look upon a piazza, and just behind it, over a low wall, there is an immense wasteland, train tracks, and a lone industrial building, the remnant of a once much larger Pirelli factory. "What is that empty land?" I ask a student who is doing research on deindustrialization. "It is the foundation of the university," he answers, referring to the fact that the university buildings have been constructed on dismissed areas previously occupied by Pirelli. More to the left, just outside my view, there is a giant movie theatre with at least half a dozen viewing halls. In a place where town centres are still important points of reference, and where multihall cinemas and large malls are a recent addition, the university, cinema, industrial premises, and the empty unused land constitute a surprising landscape.

Finding a friend in one of the massive buildings of the university gives a new direction to my journey. My companion and I walk out of U9 towards Sarca Avenue, a busy and large avenue, and go looking for Casa Loca, an independent community centre located in an (illegally) "occupied" building. "Is it far?" I ask her, as I have never gone there from "this" side of town. In my mental map of the city, the Casa Loca that is part of an older, working-class neighbourhood is miles removed from the massive new buildings of the

University of Milan-Bicocca whose names U6, U7, U16, etc. do not cease to
remind me of science fiction films, or assembly zones in giant factories. "No,"
she replies, "It is only three blocks from here."

Soon the shiny glass structures, interspersed with pieces of "industrial
archaeology" – remains of industrial equipment or parts of buildings that
have been left to commemorate an era – give way to much older, brown brick
buildings and cafés. After a few minutes, we reach our destination: a house
that has been completely repainted a dazzling sky blue, with giant Zapatistas
figures imprinted upon it. In the courtyard of Casa Loca, the sun pours over
the tables and chairs where people are eating, chatting, and laughing. No one
asks us who we are, or what we are looking for. Instead we ask the people in
the kitchen about news of Casa Loca, or of other independent oppositional
centres. The woman in the kitchen responds with a litany of names: the Bulk,
the Orso, the Garibaldi, the Stecca, the Pergola. They are all names of places
and organizations like this that since 2005 have been dismantled, dispersed,
and closed by the municipality and its right-leaning mayor. She does not tell
us much else, suggesting that this list alone is enough to depict the situation.
This time, on the door of Casa Loca there is not the protective statue of San
Precario, a satirical "invented" saint, which had been here when I first visited
the centre in 2005.

Neoliberalism has meant also a reterritorialization of the state in relation to its cities. The role of the former changed from promoting national
integration and growth to organizing conditions for an optimal capital
accumulation in those urban regions that act as key nodes for global capitalism (Brenner, 2004; 1998). To promote the growth and competitiveness
of these particular urban regions, most European countries, including
Italy, have been involved in various economic development projects, such
as training programs, science and business facilities, and urban redevelopment. This has, in part, been implemented by giving more power to
regions, provinces, and cities. In Italy, the 1990s saw the shift of many
functions – including employment services – from the state to regions
and municipalities. This decentralization was meant to encourage local
institutions to take a more active, entrepreneurial role in business development in their territory (Gualmini and Ferrera, 2004).

This can help explain why the neoliberal reshaping of Milan and its
region has gone hand in hand, at a spatial level, with an urban regeneration process aimed at strengthening the position of Milan as a key European player – both by improving its image as a modern and dynamic
metropolis and by supporting economic links and processes. In the

past decade, in fact, Milan has been involved in more than thirty major urban redevelopment projects, affecting almost 11 million square metres (Goldstein, Bonfantini, and Botti, 2007: 14). These work-in-progress zones have been creating new residences and tertiary facilities, accompanied by green spaces and new infrastructure and/or services – such as shopping malls, theatres, and/or educational buildings.

To cite some examples, among the recently completed projects there are several new areas in the Bovisa neighbourhood, the so-called Bicocca Tecnocity (where new residences, services, the University of Milan-Bicocca, and the Arcimboldi Theatre are now located) and the external Fiera exposition and convention centre. The External Fiera, which was inaugurated in 2005, is one of the largest exposition fairgrounds in Europe. Moreover, as the exposition functions and structures moved there from the previously existing, Internal Fiera, the latter became the object of restructuring. This project, called City Life, and whose completion is planned for 2014, will create new residences, a park, and a museum of contemporary art. Other current projects include Santa Giulia, the European Library, the Southern Agrarian Park, and the Garibaldi Repubblica area located very close to the Centre; see Goldstein, Bonfantini, and Botti (2007) for a description of these projects. The latter will include a new central district of fashion with a museum and educational facilities, new headquarters for the municipal and regional administrations, exclusive new residences, and tertiary facilities. In addition, Milan will host the World Exposition in 2015, and the preparation for this event includes the creation of new buildings and transportation lines.

The responses to the "new" Milan are varied and often uneasy and contradictory. On the one hand, redevelopment has generated enthusiasm and the hope of urban growth. Areas of the city that had become interstices (Tonnelat, 2008) and vacant lots, officially devoid of functions, could finally be redesigned and converted into residences and amenities. An important feature of all of these developments, in fact, is that they take the place of industrial dismissed areas (*aree dismesse* – the empty factory buildings and land that have been "left over" from the process of deindustrialization): Santa Giulia on the former Redaelli and Montedison, Bicocca on the old Pirelli, and the Bovisa in Sesto San Giovanni on the vacant lots of Falck. Harnessing the work of well-known architects and planners, and involving both municipal and private efforts, these transformations are supposed to bring the metropolis in line with other major European cities. As Bricocoli and Savoldi point out, the redevelopment zones "speak to the public of global cities with a language that is easy to decode" (2010: 197), thus raising hopes of affluence and renewed economic vigour.

On the other hand, the projects have been marred by political fights, accusations of corruption, and threats of economic downturn. Commentators have criticized the City of Milan for a lack of careful, long-term planning, and they worry about the gentrification brought by these projects (Bricocoli and Savoldi, 2010). Because the residences being built on the new areas tend to be prestigious and expensive, they are likely to contribute to the current flight from the city by the less-privileged inhabitants of Milan. What concerns several scholars, moreover, is that urban gentrification, coupled with a reduction of services and the deregulation of labour, can lead to the further marginalization of those Milanese who are not in a position to reap the benefits of Milan's dynamic economy (Bonomi, 2008). "The neoliberal city," writes Asef Bayat, "is a city shaped more by the logic of the Market than the needs of its inhabitants" (2012: 111).

I leave Bicocca to travel to another district in transformation in the north of the city: Bovisa. After a long bus ride, I reach the stop where Eliza is waiting for me with a camera in her hand. A working-class, elderly woman with a keen passion for her city, Eliza has become an invaluable walking guide for me as well as a friend. Today she is meeting me to show me this neighbourhood where she has lived all her life. It is cold, and we take turns holding our cameras to put on mittens and hats. We start walking, and Eliza describes almost street by street the places that are now invisible, what remained the same, and the new buildings that she had seen being constructed:

This house here is where I was born ... it was called the big barrack ... because there lived 110 families. It was a village. People also lived in the attics. There were cellars, where we escaped to during the war ... it was a house, but it was also a village. Then the municipality took it. They wanted to do homes for rail workers, and then they wanted to do offices ... So [...] they walled it up, until the time they would know what to do with it. They walled it [...] from 1970 to 1974. "Today we will demolish it" [they would say], "tomorrow," who knows?" So the 1990s came. At a certain point we saw the scaffolding ... they were restructuring it ... and so it lived anew. The house is more than one hundred years old because when my parents moved there in the thirties it was already old [...]

When we were children, it was not like this [...] There were gardens, and there we went to play [...] There was a radio that was bombarded during the war ... There is an enormous change. That house remained, and this one they are redeveloping it. This one is still there but as you see [is also being transformed] ... Now here is the Porsche [car dealer], and before it, there was a paper factory [...]

[This was] a telecommunication factory ... It was a post-war factory, I think that five thousand people worked there, it was a resource [for the neighbourhood] ... When they demolished it, we all kept asking ourselves: "My goodness, was it ... was it so big?" And when we would see the empty space inside, passing by with the bus, we would say to each other, "My goodness, what was inside?" It was a marvel!

With every step, Eliza tells me about her childhood, about growing up in a self-consciously left-leaning, working-class area of the city, where factories were central landmarks. When its last one, the Face, was demolished at the beginning of the 1990s, impressing one last time its vastness and centrality on the neighbourhood, Bovisa was already on its way to becoming a very different place. Revisiting these spaces, most of which have shrunk or disappeared, prompts Eliza to reflect on wider shifts in the identity and lifestyle of people living in Bovisa. In earlier times, she explains, relationships were built on everyday neighbourhood contacts. Instead, "now everyone has different interests," which serve as stronger reference points than the identity of the area. "Think of the big barrack," she reflects, "If you needed something, there were a hundred people to help you out." And pointing to the section of street we had just walked on together she notes:

This section here, where we just walked – now we did not meet anybody who I said good morning to – Then, at each step, "Hallo! Hallo!" There was a strong community that now is not there anymore. It was not just at the level of one building, it was – how to say it – on people's skin.

Eliza's itinerary, however, does not stop here. She also shows me, with a sense of enthusiasm and curiosity, the new Bovisa, the one brought by the university (the recently established new section of the Politecnico), the Triennale art centre, the Mario Negri research institute, and the new businesses. These are all "beautiful" new buildings, introducing novel activities, people, and sights into this area. With the guidance of her commentary and her camera, I start to see the transformations as a dialogue of textures: the rough, opaque, and earthy surfaces of the old have been giving way to shiny buildings, and smooth, curved glass outlines that convey a sense of rupture and futurism.

Yet as we approach the first one of these sparkling new edifices, the Bodio Centre, our walk and narration take an unexpected turn. While Eliza snaps pictures of the curved roof and its courtyard accessible through open gates, and I follow her with my video camera, a man in uniform rushes to us from inside the building. "You are not allowed to photograph!" he informs us curtly. We hesitate, not wanting to end so abruptly our visit, yet the young man makes sure that we cross the street and go elsewhere. Our struggle with surveillance cameras and security guards continues at several other buildings, including the university, where we are told that we need to see the authorities in order to be able to take pictures of the buildings and courtyards. The feeling that we are being controlled while trying to witness these transformations becomes visible in Eliza's discomfort at having to play spy in the streets and piazzas that she has known so intimately since the time of her childhood and of which she can narrate so exactly the stories and events – much more exactly than the people who now guard them so carefully.

We continue our walk, and with the precision of a mapmaker Eliza's narration equally takes hold of all parts of the neighbourhood – from the new tunnels covered in graffiti, to the older churches and piazzas sitting side by side with empty fields, and massive new buildings made of glass. I could say that we are simply making an inventory of landmarks and places. But, as Tsing (2005) warns us, there are always complex relations at work in list making. Here, our walk is becoming a way to construct a location from where Eliza can tell me about the area and its transformations as a subject in space who can anchor the past and the future. From this point of view, however, the movement of time is less the succession of buildings and forms, than a simultaneous appearance of the old and the new in the streets and landscapes of the neighbourhood.

As we reach the new train station, Eliza and I start to note with an eerie surprise that the metallic frame of the "gasometer" (a very large industrial structure which was used from the early 1900s to store gas) that has been left as a historic landmark, is visible from almost all areas of the neighbourhood. Its skeletal frame seems to follow us around, and is casting its shadow on abandoned lots as well as on shiny new buildings. "It is the spirit of the place," exclaims Eliza. She does not say whether it is more akin to a spirit's blessing or a ghostly haunting. In many ways, in this neighbourhood still in the process of becoming, it is this very difference that is at play.

Neoliberal Milan has been marked by a growing focus on security, understood as personal safety, and disassociated from wider, collective notions of social and existential well-being (Paone, 2008). This mirrors wider discourses and concerns: the perceived need for more security has been a dominant theme in contemporary Italian national politics and a key legitimizing factor for stricter anti-immigration laws (Quassoli, 2004; Dal Lago, 1999). Here it is very important to note that refugees, Roma and Sinti persons (whether Italian citizens or not), and immigrants (especially those who do not have legal visas and permits) have increasingly taken on the role of the enemy and the main threat to public security.

This has been instrumental for rethinking and reorganizing social relations and structures. More particularly, programs meant to integrate underprivileged groups and to redistribute wealth have been largely replaced by repressive policies (Saitta, 2011:112). Among the latter are violent reactions against dissent, spatial governmentality measures, and the "hyper-incarceration" of marginalized populations (De Giorgi, 2010; see also Isin and Rygiel, 2007). Indeed, for the latter, detention in its various modalities seems to be the last remaining and only available form of state "support" (Melossi, 2003: 378; Dal Lago, 1999). Neoliberal deregulations and the growing precariousness of labour is, then, the other side of the coin of an increasingly repressive state. Describing this situation, Loïc Waquant labels the new regime "liberal-paternalist," because it shows a liberal disposition to the players in the most privileged sectors of society, while being "paternalistic and punitive at the bottom, towards those destabilized by the conjoint restructuring of employment and withering away of the welfare state protection or their reconversion into instruments of surveillance for the poor" (2001: 402). At an urban level, a focus on security has been a crucial factor in the municipal government of Milan adopting a "zero tolerance" framework, modelled after Mayor Rudy Giuliani's efforts in New York (Quassoli, 2004; Waquant, 2001). This approach sees repressive policies as the most effective tool to counter perceived insecurity in the city. The summary of the initiatives carried out in Milan by the 2006–11 Mayor Letizia Moratti, contained in a pamphlet distributed by the latter to all households just before the municipal elections in 2011, is telling in this regard. It explains that "in order to counter the problem of the areas of Milan with a high presence of immigrants, the municipality has put into effect a multilevel strategy, which strengthens security as well as integration" (Committee for Letizia Moratti Mayor of Milan, 2011: 72). All the initiatives listed under these words, however, refer to police checks, arrests, evacuations, and the diminishing of populations seen as problematic. There is only one note about a support program, called "a journey of recovery," but no details are given. Not surprisingly, in my conversations in 2009 and 2011 with people involved in immigrants' rights organizations, my interlocutors characterized the government of this past decade as being in a war against disenfranchised communities. Among the latter are Roma and Sinti, who have become the targets of "emergency" measures in various regions of Italy (Sigona, 2011). Significantly, the Moratti committee's electoral booklet indicates that to improve security in the city, the number of Roma and Sinti living in Milan and its surrounding area has been reduced by 81% between 2007 and 2010 (73).

It is worth pointing out here that, in this context, security could easily signify a very different set of ideas. The discourse of concern for security in relation to Roma/Sinti settlements could mean to increase the safety of people living there, for they are often the targets of arson and xenophobic attacks. Or it could lead to imagining ways to guarantee social security, such as job opportunities, welfare benefits, and inclusion in society. On the contrary, when security is advocated in this context, it only refers to non-Roma/Sinti Milanese inhabitants, who see nomadic people and migrants as a possible threat to *their* living conditions. Moratti and her party, moreover, introduced important by-laws to curb drug use, panhandling, graffiti, and street selling. Many inhabitants welcomed these initiatives, but many others bitterly resented tighter regulations and police control.

Politically, the Moratti administration has been part of the long rule in Milan by right and centre-right forces led by Berlusconi and the Northern League – who dominated the regional government, too, since 1995. This political constellation came to power after one of the landmark events in the recent history of Milan: the discovery of *Tangentopoli* (bribes-town), the extensive system of bribes that had been sustaining the majority of the political élite. This scandal, which erupted in Milan in 1992, but then spread to many other locations in Italy, ended the rule by the Socialists and the Christian Democrats that had marked Milan from the 1970s. In its place, the secessionist Northern League came into power, with an anti-immigrant stance, a focus on reorganizing the government, and a disdain for the Milanese Social Centres movement.

From the time of the Northern League and until the last (2011) municipal elections, right and centre-right coalitions have led the city (with arguably no overwhelming change in the matter of endemic corruption; see also Guano, 2008). Berlusconi, in particular, has been very influential in Milan, both as a very wealthy businessman – who has built entire neighbourhoods in Milan and who controls much of the media – and as the political leader of Forza Italia, a political party that John Foot describes as a "post-modern populism" (2001: 183). In this past decade and a half, Milan could indeed qualify as "the city" of Berlusconi. It is here that he has built a great part of his financial empire, and it is from Milan that he launched his political movement. Notwithstanding his stronghold, in spring 2011, a landmark municipal election challenged his rule and brought a centre-left coalition led by Giuliano Pisapia to power.

Leaving Eliza, and returning from Bovisa, I reach the much more central part of town where I am staying. It is a somewhat incongruous neighbourhood: within a few blocks you can find the headquarters of two of the most prestigious fashion houses in town, several large social housing projects, and another one of the many dismissed areas, this time the remnant of an abandoned train station. Tired from my journey, I sit on a small bench under a tree, close to the streetcar tracks. It is at this very spot that I have met a few times a very old woman sitting and waiting for her bus. She has grey hair and very blue eyes. She is usually dressed in layers, with a long simple black skirt, a very modest dark blouse, and an ancient dark blue sweater. Her simple clothes and the glaring absence of accessories, bags, or jewellery render her strikingly out of place in this part of town, between people who are impeccably elegant. The last time I was here, she smiled at me, a warm smile that showed a mouth where many teeth are missing, and asked me about my child, a two-year-old who was waiting with me. "They are tiring, the children. They made me tired," she said. "Oh yes, they are a lot of work. I have two of them, and they make me tired, too," I replied. "They tired me," she repeated, "I had ten." "Ten? No wonder they tired you!" I said. She told me about her children, about playing with them in her small town in the south of Italy, about cooking for them, about doing laundry for all of them by hand. She looked at me intently with her brilliant blue eyes that contrast so strongly with her black layered clothing – as much as her presence and demeanour contrasts with this neighbourhood brimming with boutiques, with bars and their fashionable drinks, the passers-by with Prada boots and stylish handbags. And I wondered: how does it happen that the boots and the drinks render someone out of synch with their surroundings, rather than they themselves becoming strange and unfamiliar? How did instead the blue-eyed grandmother become a "humanity in excess" [Paone, 2008: 85], almost a time traveller in the very streets she inhabits?

The woman with very blue eyes is once more a reminder of the coming together of different voices and trajectories. In many ways, Milan is a city of paradoxes and disconnections. How can it be, for example, that in a city known internationally for its elegance and textile production, on three occasions in a six-month period (from October 2004 to March 2005) a person died trapped in a charity collection bin while trying to take some used clothes from it? How can it be that, while the lack of affordable housing pushes many inhabitants into poverty, hundreds of apartments – or even entire buildings – remain empty and unused, sometimes for many years (TempoRiuso, 2009; Multiplicity.lab, 2007)?

Scholars describe Milan as an incomplete metropolis, traversed by gaps and pieced together by contradictions – a city of "shards" (Bonomi, 2008), where different parts of the city and sectors of the population often coexist without dialogue or mutual engagement. They call it divided, suffering, difficult, and fragmented (Goldstein and Bonfantini, 2007; Multiplicity.lab, 2007; Petrillo, 2004; and Foot, 2001). What commentators mean with these phrases is that, due to the many, often disorienting, changes in Milan in the past sixty years, history is felt and understood differently in each neighbourhood, and its populations have dissimilar memories, priorities, and even nostalgias to refer to.

In this context, fragmentation is itself a way in which the city hangs together and creates a social terrain, rather than what simply separates it into distinct entities. This situation, moreover, is not limited to Milan. It is rather a condition generated by contemporary global processes (Bayat, 2012; Paone, 2008; Drieskens et al., 2007; Smart 2003; Caldeira, 2000). In turn, these discrepancies make it difficult not only to understand but also to represent Milanese realities. Stefano Boeri, for example, suggests that standard sociological portraits of Milan have missed some of its key characteristics, very significant circulating ideas, and the underlying "rhythm that unites places" (2007: 9). This is because, by describing well-defined, larger forces and structures, they failed to address the more minute, heterogeneous, and at times opposing, "kaleidoscopic" activities constantly shaping and reorganizing the city.[12] Similarly, one of the questions that confronted me during my research has been how to approach, analyse, and write about a social reality that very often seems composed of echoes rather than clear, definite melodies. The narrative juxtapositions I use in this and other chapters are a strategy to try and represent this jarring coming together and apart of aspects of the social.

Chapter Three

The Agora of the City

At the end of December 2004 a small yet inspiring event caught many Milanese by surprise. A new citizens' movement emerged almost overnight, with the goal of fixing some of the problems of the city not through the available institutions, non-profit associations, or political organizations, but directly, through the grassroots, concerted efforts of regular citizens and neighbours. Existing at first as an Internet site and a blog, it soon became an active presence in the city. Interestingly, the first public event of this movement, which was calling itself *VivereMilano* (Living Milan), was to hold an open street meeting, in January 2005, in one of the most prominent public spaces of the centre of Milan: the Galleria Vittorio Emanuele. Assembling many residents who had never taken part in demonstrations before, and generating a sense of enthusiasm and hopeful optimism in both participants and observers, this forum was a striking example of a grassroots deployment of public space for efforts geared towards social change.

VivereMilano grew larger in the following months, organized several events and campaigns, and has been active throughout the years since its founding. Here, however, I write about this first meeting, because I am interested in how this movement took hold of the piazza – both materially and imaginatively. As I followed the birth of the movement in the streets, I found myself asking: How did VivereMilano use the piazza for its purposes? What understandings of public space did it foreground, and how did this affect what VivereMilano could say and do?

VivereMilano

The street meeting of VivereMilano, which was the official birth of the movement, happened approximately a month after the *Corriere della Sera*, one of Italy's leading newspapers, published a letter by Cesare Fracca, a professional man in his late thirties (Schiavi, 2004). In his letter, Fracca lamented the poor quality of life in Milan, and he urged other people of his generation who, like him, had always been disinterested in the problems of their city, to wake up, and to do something about the deplorable urban situation at hand. The letter provoked a landslide of responses by people whose experience of Milan resonated with Fracca's. With the support of the newspaper (particularly of the editorial section that focuses on Milanese news), Fracca and his correspondents started a blog – an Internet site where people could post articles, comments, and letters on the city and its problems, and devise possible solutions.[1] VivereMilano quickly expanded as more and more people visited, read, and wrote on the site. In this process, Fracca continued to play an important role, as one of its most avid bloggers, informal leaders, and organizers.

At the beginning of its life, because its participants had been only reading about and writing to each other, VivereMilano was confronted with the problem of how to make itself "visible" in the actual streets and piazzas of Milan. An interesting way it sought to do this was to choose a colour for itself: orange. As some of the bloggers suggested, if the people who are writing and/or reading on the Internet site could wear something orange when they are in the city (e.g., in piazzas, streets, open markets, and while using public transportation), they might be able to recognize each other. This would serve to mark the movement's existence to outsiders, too.

VivereMilano called for an open meeting to discuss the ideas and issues that emerged from the letters and the blog postings. This gathering took place on 23 January 2005, a very cold Sunday morning, at one end of the Galleria Vittorio Emanuele, where it leads into the Duomo Piazza. Open to passers-by and to anyone who would bother listening and/or talking, the meeting had a strongly informal character. About 130 people congregated in a circle for approximately two hours, and people expressed opinions and asked questions without loudspeakers, microphones, or a podium. People circulated along the rim of the circle, meeting each other, conversing with strangers and with people they already knew, and engaging curious passers-by. A manifesto of the movement had been prepared by Fracca and a small committee that had met some time earlier, and it was handed out to anyone who cared to take it.

In writing about this meeting, I am particularly interested in the way in which VivereMilano made itself present in public space. I am intrigued by how the movement evoked the ideal of public space as an arena of debate and of confrontation, all while closing off and pushing to the margins certain topics and avenues of conversation. The forum, in fact, brought up many problems and aspects of daily life in the city, including air pollution, traffic, public transit, housing prices, the role of the municipal government, and more. As interesting as what the participants say, however, is what their comments do not consider, such as the issues and difficulties encountered by poorer residents, by people of colour, and/or by immigrants. The way in which this group brings, literally, certain topics and certain exclusions into public space can offer us a perspective on some of the dynamics and complexities of public space in Milan and help us reflect on who might be able to claim a legitimate presence in city spaces.

The participants of VivereMilano are in a relatively privileged position, but this should not blind us to how their generation (the cohort of 30- to 40-year-olds) is significantly affected by the increase of precarious working conditions and the decrease of social and economic security (see also Molé 2010). Significantly, while 30- to 40-year-olds have a very low poverty rate compared with other age groups, Benassi (2005) explains that this is partly because many of them are still living with their parents, even if they are employed. In his research, Benassi found that half of the 25- to 34-year-olds (three-quarters of whom were working) had not been able to establish a household of their own. "Waking up" to a leadership role towards their city, as VivereMilano members attempt to do, might be a way to try and overcome this difficult situation. However, it is very interesting that none of the participants in the forum responded to a participating politician's comment that the 30- to 40-year-olds are part of a generation in transition.

Here below I present three excerpts from the tape recording of the forum. I quote the discussion at length in an effort to represent as best as possible some of the variety of the themes and the multiple directions of the conversation. I gave pseudonyms to speakers in order to make the discussion easier to follow, but would like to remind the reader that most of the people who participated in the meeting did not know each other's names. I use theatrical stage directions to convey the setting, and the role played by affect, voice, humour, and staging. Those aspects were, in fact, an important part of what the forum was trying to accomplish and of the way it was, literally, *taking place* in the space of the Galleria.

Stage directions are not usually part of ethnographies; however, several authors including Bridgman (2006), Fabian (1990), and Madison (1999) have used forms inspired by theatre and performance studies to write academic texts and to discuss social theory. Johannes Fabian, in particular, argues that if we simply focus on the content of social interactions, we might "fail to account for historically contingent creation of information *in and through the events* in which messages are said to be transmitted" (1990: 11, emphasis in original). In relation to this meeting, not only what was said was important, but also where it was said, and how participants positioned themselves as speakers in space. It is important here to note that a performance of informality (the circle of participants without leaders or microphones, emulating a regular conversation between a group of friends) was an integral part of the enactment of public space as a symbolic agora – a central piazza where all city residents can gather to speak freely (notwithstanding that the Greek agora was limited to men and non-slaves).

As I do in chapter 8, I wrote myself as one of the characters in the "play." In doing this, I wanted to emphasize that this meeting was listened to and recorded from a particular position, and not from a distant and encompassing vantage point. On the one hand, this means that it is necessarily a partial representation: during the forum I was part of the group of participants, and I certainly missed some of the comments that were spoken in softer voices or that came from more distant parts of the circle. On the other hand, I seek to emphasize the role of my situated and embodied trajectories in the city for crystallizing fleeting moments and encounters into a text and an academically grounded reflection. To say it simply, my stepping into and out of the circle of the participants is what enabled particular stories, issues, and meanings of public space to emerge.

THE FORUM

I.

(Setting: A group of people (approximately one hundred) is gathering by one of the big stone pillars of a promenade with arches. The Duomo cathedral and the steps leading to it are visible just behind the promenade. On the pillar, placed higher than the gathering people, are a street sign and a stone inscription. The gathering people are forming a circle and are taking turns talking – sometimes quietly, sometimes very animatedly, with frequent laughter, interruption, and overlap. There is a sense of excitement and celebration

among them. It is very cold: many people are rubbing their hands and shifting from foot to foot to keep warm, and they can see their breath coming from their mouths. They have hats, mittens, and winter coats on, and almost all of them are wearing something orange (like a scarf, gloves, or a jacket). A young man is wearing an orange full bodysuit. Many people are chatting on the periphery of the circle. Some others are moving around it, trying to find a good spot from where to listen and see what is going on. Passers-by are walking by in all directions. Some join the circle, some continue on. Two people are distributing a pamphlet with the manifesto of the group. The people assembled in the circle take the paper eagerly, while some of the passers-by are interested in it and some are not.

Cristina/the anthropologist approaches the circle, takes the pamphlet and starts to read it.)

> *A new sentiment is circulating in the streets of Milan:*
> *it is the spirit, made of passion and awareness,*
> *of us awakened citizens.*
> *VivereMilano is a spontaneous movement by 30- to 40-year-olds*
> *who stand up to show that they exist,*
> *and that they are not alone*
> *in their wanting to design and to build a better city.*
> *VivereMilano is a laboratory without prejudices nor alignments,*
> *free to develop ideas, projects, and concrete actions,*
> *which acts as an [...] amplifier of messages and signals.*
> *[...]*
> *In our life we have cared very little for "the public thing": we always preferred that others engaged in politics, because it appeared boring, at first sight tiring, and ineffective.*
>
> *Our strategy has always been the escape: our goal was to work as much as possible, holding our breath for five days in order to re-emerge from the apnea for the brief diversion of the weekend.*
>
> *Today, however, something in us is changed, and we realize that the quality of our life and of our children is totally influenced by our disinterest. Finally we wake up from our apathy, and we start to talk about dreams, ideals, and concrete possibilities for a better life. We want to start to reflect also with our hearts and our bellies, distancing ourselves from the empty logics which characterize the politics that we saw and we voted until now; we want to untie ourselves from parties and give clear contents to politics, which is today distant from us and little convincing [...]*
> *[VivereMilano, 2005]*

(Cristina/the anthropologist puts the paper in her bag and joins the circle to listen to what the gathering people are saying.)

ARTURO: (*in a nostalgic voice*) [...] twenty or thirty years ago [in Milan there were more] relationships, [a feeling of] human belonging ... Milan was not always the same [as it is now]. There was a ... there was a tradition of welcome, of solidarity [...] It was a place where there was work [...] It is a city that welcomed generations of people that arrived, that gave them work [...] This according to me gave a sense of *belonging* to the city, which now slowly has been lost, lost for thousands of reasons [...] [Even] La Scala [theatre] [...] once had a different value ... the people who went to La Scala stayed here on the weekend, they met [...] they *talked* about this city, while now they go after their own interests.

[...]

UGO: This thing is getting bigger, and is happening also in other cities [...] a movement similar to this was born in Napoli. They have put together the professional associations and they held a conference in [a] [...] theatre of the city [...] A movement is emerging of people who never cared about politics but who have high ideals [...] who after years of seeing their problems not respected, decide to solve them in the first person. For me, it is the first time not only that I participate in a demonstration, it is also the first time that I speak out (*laughs*) [...] I believe that this is the important thing: to start from the problems of the city [...]

DANTE: (*insistent*) I wanted to make my proposal! I wanted to make my proposal!

UGO: (*interrupting the previous speaker*) for the droppings of dogs, for example, how many people do really –

DANTE: (*annoyed*) I wanted to make my proposal!!!

(Several voices start discussing in the background.)

LEONARDO: Let's not blame the municipality for this [the droppings of dogs]!

ARIANNA: Can I say something so that I can go to mass since it's 11 o'clock? –

UGO: (*interrupting the previous speaker*) Go to the 6 o'clock one!

ARIANNA: I don't know if this is too big a problem, if it is another topic, but the *costs* of this city, for example, the houses –

GAIA: (*interrupting the previous speaker; with great emphasis*) Oh, great!!

(A man starts talking about housing prices in the background.)

ARIANNA: This is something that really concerns me, buying a house means –

LUISA: (*interrupting the previous speaker*) draining all blood from your veins [i.e., going bankrupt]

ARIANNA: It is a big problem … to access a mortgage now I should prosti-
tute myself …
GAIA: (*loudly, emphatically*) Well said! Me too! If you want, we can go together

(*Laughter. Man talking about houses in the background stops speaking.*)

ARIANNA: (*with resignation*) Too bad for us, for the 30-, 35-, 40-year-olds
who want to buy a house … Just that.
[…]
ETTORE: Can I make a proposal of method? Because I believe that if we
talk about air pollution we *all* agree, [if we talk about] noise levels, we
all [agree], [if we talk about] socialization that is lacking, we *all* [agree],
otherwise we would not be here. Moreover, I would like to point out that
we are more than 130, for this I thank you […]
Regarding the method: I seem to see two things. The first [focus we could
adopt] is this: the bicycles, the droppings of dogs, etc. […] The second is a different city. According to me, between these two points there is a long journey,
but […] most of all there is our direction, that is not to play municipality two,
the municipality for the bicycles, for the droppings of dogs; we have no time
nor sensibility to do that […]
Our concrete perspective […] could be, first of all, even if banal, finding a physical place that is warm where we can talk, […] secondly, finding
a route that would not be micro nor macro […] I will give some random
examples:
– we could either reflect according to groups of citizens: the elderly, the
kids, those who do not have spaces […]
– the other could be to reflect like persons:
persons have […] a sense of smell, so we do not want to breathe disgusting
dirt; a sense of touch, and we do not want to step on [dogs'] excrements
ANTONELLA: (*interjecting*) a sense of sight
ETTORE: a sense of sight, and we do not want to see monsters, a sense of
hearing, and we do not want to hear disco noises all night long, and this
[direction] could be, if you want, another way of getting together […]

II.

FIRST POLITICIAN: […] The first thing that I wanted to say was that I wanted to thank the one who has thought [about this meeting] because the
fact that today, in Milan, the Duomo Piazza has returned to be the agora
of the city – that is, the place where the citizens meet and discuss the

problems of the city – is an important reconquest of a piece of democracy […] There is a further topic that I want to bring up […], the problem of precarious working conditions, and of work. I think that this generation of 30- to 40-year-olds has one more problem than the one of the quality of life, or of housing, etc. It is the one of finding a role, and a social identity despite the difficulty of not finding any more the securities […] that the generation of our fathers or grandfathers had […]

UGO: And you think that this can be solved at the level of the city?

FIRST POLITICIAN: No, absolutely, it cannot be resolved only at the city level. But

(Several voices interrupting, discussing.)

FIRST POLITICIAN: Excuse me, I conclude. But […] [action at the city level] can help look for guarantees and welfare systems that could resolve some of the problems caused by precariousness […]

[…]

UGO: […] Now only 20 days have passed from our birth, but […] what is giving us enthusiasm is that this participation that is widening, from below, from people who have never done politics […] could become something more.

[…]

DONATO: Excuse me, my name is [Donato] […], I am 42 years old. I only heard right things/good points. There is one point on which I am perplexed, and that is the one […] of keeping the distance from the Palace [the site of institutional politics].

TANIA: *(interrupts the previous speaker)* Exactly!

DONATO: We are different; we are outside of the logics of centre, right … I understand all this very well, but there is, so to speak, a contradiction … which I have noticed […]: it is not that the government/administration is assigned to us through a *lottery*. Then we make a list of things … we bring them there … and if we were lucky in the lottery [we obtain what we need] […]

[Municipal politics] is a mechanism that has to be kept under control, it is a mechanism by which … there must be a *link* between 130 people *(laughing)* who meet [in the street] and the ones who go there [the politicians] –

UGO: *(interrupts the previous speaker)* Can I say one thing? […] I see that in many years

DONATO: *(interrupts his interrupter; very animatedly)* they are not all the *same*, the administrators [city councillors]! If we have terrible ones it is *our* fault! It is not that we were unlucky at the lottery!

[...]

UGO: Each one of us has voted right, left, centre –

VERA (*interjects*) without being interested!

UGO: Without being convinced, being less and less convinced, at times voting against –

FEDERICO: (*interrupts the previous speaker*) So let's not vote!

UGO: We go here so that we do not end up there [...] I have never voted really convinced [of my choice], I say ... I vote this because it is the least of the evil.

TOMMASO: (*interrupts*) So, listen, if it is like this, I prefer that one like you, in the end, [...] would go himself to make politics [...]

MARCO: I believe it is still too early to come to this!

TONINO: We are already making politics ...

AMATO: The politics of institutions!

UGO: But, I say, let's not feel guilty if we have elected [...] and we are absolutely not against this [current] administration ... we are against ... let's say we are for an improvement ... the administration, if it has not represented us, it is not because we chose the wrong person, but perhaps because we chose the person who ...

TIZIANA: (*interrupts the previous speaker*) [...] has represented other interests much bigger than ours!

GIULIA: (*interrupts the previous speaker*) No, but apart from the interests, she or he has not wanted to listen

[...]

MARCELLO: Can I intervene, excuse me? [...] I seem to understand that we could be born as a pressure group, that is, outside of politics, yet inside –

THEA: (*interrupts the previous speaker*) It is like this, it is like this!

MARCELLO: without a colour [association] which is really political, but we can press for politics to *orient* itself in a definite direction [...] because if we enter in the mechanisms of politics, oh no, we are finished. But at the same time, it is necessary to interact clearly with politics, because obviously politics is an instrument that is at the basis of all decisions, of the city, etc. Then there is a thing I would like to say [...]: the people who vote, many vote without knowing who they are voting for, [...] they vote simply because they *have* to, they vote simply because they saw a commercial on TV. But very often they do *not know* really who they are voting for –

FORTUNATO: (*interrupts the previous speaker*) Well, now, it is not that ...

MARCELLO: (*very animatedly*) Look, many people are like that, guys, many people are like that

UMBERTO: (*interrupts the previous speaker; annoyed*) Who cares, lets talk about problems!

(Many voices discussing.)

[…]

MARCELLO: *(agitated)* Democracy! Democracy is when somebody knows what they are going to do! Here there are people who do not know what they are doing! That is the reality!

(Many voices discussing animatedly.)

MARCELLO: in fact, we are here in 130 and we are not even capable of –

UMBERTO: *(interrupts the previous speaker; annoyed)* we are not doing a debate on the electoral system! Let's talk about the problems of the city! Let's talk about the problems of the city, we are not here to make a system, a new electoral system …

[…]

SECOND POLITICIAN: *(talking really rapidly)* […] You already have a colour, a very beautiful one, and it is orange …

GIOVANNI: *(interrupts the previous speaker)* Yes, we chose it exactly because [it does not correspond to any Italian political party colours] –

SECOND POLITICIAN: *(interrupting his interrupter)* And you did well, you did well! […] You are a small piece of the energy of civil society, because you identified yourself with the issues of daily life, that is, the urban malaise […] But maintain your autonomy […] This is the true democracy […]

[…]

THIRD POLITICIAN: […] I came here just to listen […] I want to let you know that the doors of my office are open […]

III.

UGO: We want to put together as many people as possible […] not to compile a list of complaints, but to compile a list of solutions to problems […] We talked about children, about pollution, about dogs' excrements, about graffiti, about public transportation, etc. etc., but perhaps there is something else, I am sure. These are the topics we have woken up around. The intention is to listen to all, […] and then […] to propose concrete measures. For this we also need competent persons, I repeat we are persons, we are people like you, who have never been involved in politics […] I invite lawyers and doctors, we need doctors, to join us, […] and then to propose solutions […]

[…]

CORRADO: I want to say something very banal. The most important, and banal, thing is to be able to *live* the city. We talked about the example of Rome, that, although it has problems [...] that are more serious than the ones of Milan, they [people in Rome] are able to live the city. We escape on the weekend: in the winter to the mountains, in summer to the seaside; the most important thing would be to really live the city ...

[...]

PIERPAOLO: It is hard, however, to promenade in Duomo [square] ... in other cities [...] people go to Piazza Navona [a major piazza in Rome] ... on the weekend I escape to [Tuscany], I do not go promenading in Duomo [...]

[...]

MATTEO: Another thing that existed years ago was the courtyard [of buildings] where kids used to play

PIETRO: (*interjecting*) Now it is a parking lot

UGO: Now it is no more [...] we would like to propose to reactivate the courtyards

CARLO: (*interjecting*) Certainly

UGO: where kids, from 3 to 5 p.m., can go play, knowing each other in the buildings/condominiums also prevents [situations like] the senior from the third floor dying and [his (her) corpse] remaining there for three months[2] ...

ALESSANDRO: Neighbourhood issues ... popular housing, that are abandoned ... the elevators do not work, the elderly have to walk up ... they are practically abandoned

ITALO: (*while the previous speaker continues in the background*) Yes, but ... But we cannot think about all of the problems of the world, guys ...

[...]

GIUSEPPE: [...] I do not know statistical data on it, but I personally also feel the problem of precariousness [unregulated, temporary, poor working conditions] [...] it is a problem ... also regarding public order and cleanliness. We talked about cleanliness, about dogs' excrements, but I am also annoyed by the people who illegally ... sell, the people who beg ...

GUIDO: [...] I would like to point out a common aspect of all the complaints that have been made [...] that is the *stu-pi-di-ty*. That is, we in Milan live in a way that is absolutely stupid! (*Starts raising his voice*) It is stupid to work eight hours a day like crazy (*is almost shouting*) to buy a house that costs three times [what it should cost].

(Several people start to applaud.)

GUIDO: and as soon as Saturday comes

(End applause.)

GUIDO: *(shouting)* we go away [to the countryside, mountains, or seaside] because we cannot live in the city where we bought a house that costs three times [what it should]!

(Laughter.)

MANY VOICES: well said!

The forum of VivereMilano exemplifies, and at the same time complicates, scholarly debates about the importance of public space and about some of the dilemmas associated with it. On the one hand, VivereMilano's act of taking hold of a central piazza to talk about the problems of the city is a witness to the importance of public spaces for creative community actions, social debates, and participatory democracy. On the other hand, the meeting shows that both spaces and publics are continuously made and negotiated in a dynamic interplay with categories such as class, race, gender, and age (De Koning, 2009; Iveson, 2007; Low, 2000; Holston and Appadurai, 1999).

These considerations are important for understanding the forum of VivereMilano. The group's work of claiming a central public space in order to debate the problems of Milan and to criticize the local government, gave the meeting its particular appeal and strong sense of enthusiasm. More, in particular, through the blog and the forum, participants expressed a sense that they were activating a potential that they saw present in public space. The fact that anyone could have joined the group, to talk about his or her concerns, suggested that Milanese streets and piazzas are, so to speak, there for the taking, as a "space within which political movements can organize and expand into wider arenas" (Mitchell, 1995: 115). As one of the politicians present at the meeting pointed out, the informal and interactive forum of VivereMilano amounted to the recreation and revaluing of the piazza as "the agora of the city."

At the same time, however, the very ease by which VivereMilano could assemble and constitute an audience in the Galleria questions this very ideal as well as the supposed accessibility of this urban locale to diverse inhabitants of the city. It becomes crucial in this respect to ask who was gathering and constituting the agora, and why, and how this very meaning and enactment of public space are linked to the particular social positions of the people taking hold of it. As the excerpt above shows, the topics discussed in the forum ranged widely, from issues about traffic and mobility, to (the lack of) feelings of pride and love towards Milan, to specific problems, events, and regulations (such as most condominiums' rules prohibiting children to play in the courtyards). One of the major concerns voiced by the participants – and, in a way, serving to unify the group – was the quality of urban life, and more specifically, the quality of the city as a sensory environment.

Indeed, both on the webpage and in the meeting, the cleanliness of city places, urban decor, graffiti, and air pollution were among the most importantly debated topics. Air pollution, in particular, was a major concern during the meeting. This is hardly surprising, considering that at the time pollutants in the air continuously surpassed the European Union's health standards and that the media were reporting on how this was thought to contribute to thousands of deaths annually. Differently from other groups more centred on environmental perspectives, however (and thus also committed to ideas such as reducing consumption, using public transportation, farming organically, and promoting environmental justice), VivereMilano's critique of air pollution and of the municipal approach on this matter seemed formulated as a part of their right to a pleasant public space.

This (although not the only) focus of the group on the sensory quality of life in the city reveals both some of the strengths and some of the problematic aspects of VivereMilano at this first meeting. One of the unique characteristics of this movement is its effort to see the city holistically, to link different aspects of daily life in Milan, and to refuse to be compartmentalized in their thinking and actions. It is for this reason that its initiators wanted to keep it apart from other, existing, organizations that focus on single neighbourhoods or problems or on particular, and limited, areas of urban life. It is a humanistic project seeking to address the totality of life and social relations in the city, and reinstating the idea of happiness as a goal and a framework for its actions. In this respect, VivereMilano can be interpreted as potentially challenging Milan's alleged focus on work, status, and allure, as well as countering the social isolation that many VivereMilano participants talk about.

At the same time, VivereMilano's holistic concern for the city and its attitude to some of the problems of Milan underscores how very middle class this movement is, in both composition and outlook. This in itself is not a problem – cannot middle-class people express critiques and use public space to form a movement? However, it is important to ask about the unintended consequences this might bring. Studies about gentrification, for example, suggest that often the interests and demands of relatively affluent residents result in the displacement of those who have fewer resources (Herzfeld, 2009; Potuoğlu-Cook, 2006; Lees, 2008). The issue of a healthy, sustainable, and green city is a case in point. As Checker (2011) argues in relation to the New York City area, the very notions of sustainability and environmental improvement tend to aid the gentrification of neighbourhoods as they are often used as tools for promoting cities and attracting higher-class residents.

We can say in this respect that VivereMilano's very focus on age (it defines itself as the movement of the 30- to 40-year-olds) as a uniting factor serves to conceal the role of class as well as of other social positions. In other words, it tends to assume that most 30- to 40-year-olds in Milan face similar challenges (while at the same time completely ignoring that those challenges might result from wider economic and social changes that affect their generation in particular ways). Although on the website and at the meeting some people did mention working conditions and the cost of living as urban problems, there was little or no talk about homelessness, poverty, racism and discrimination, homophobia, immigration issues, and the difficult conditions of elderly persons, especially single elderly women.

Several statements during the meeting were especially revealing in this regard. Although a woman raised the problem of finding affordable housing, generating an echo of consensus, the issue was framed as a mortgage difficulty rather than as a shortage of affordable (including social) housing, or in terms of the price of renting an apartment in the city. Interestingly, her joking comment about prostituting herself to buy a house points to how women, who are often working part-time or in feminized, low-paid positions have limited options, outside of marriage, to become independent of their families and to establish a home of their own. (And thus that many of them might really have to resort to prostitution in order to survive in Milan!) Indeed, gender differences between 30- to 40-year-olds were never specifically addressed, except when another woman talked about the challenges of being a mother in the city – referring, however, only to the issue of navigating a stroller through parked cars and non-accessible sidewalk passages.

Another interesting comment was one made by a man denouncing the precarious working conditions widespread in the city. Although he

might have meant to say that poverty, and the difficulty of finding a regular, long-term job, affects everyone, it is striking how in his words poverty
becomes a disturbing sight and a problem for the urban environment as a
pleasant, civilized, and middle-class space. The reference by several participants to the weekend flight from the city is also typical of middle- and
upper-class residents. As one of the interlocutors explained:

> on the weekend I escape to [Tuscany], I do not go promenading in Duomo.

The talk about the Saturday flight from the city was recurring at the
meeting and is especially interesting given that VivereMilano was born
from an explicit refusal to leave Milan on weekends – both physically
and metaphorically – in order to start engaging with its problems as
"awakened" and "aware" citizens. On the one hand, this call, also made
explicit in the manifesto, reveals how, as a matter of course, at least
some of its participants can afford to leave the city on a regular basis
to spend some time in the mountains or at the seaside.[3] On the other
hand, the conscious decision not to go somewhere else on Saturday and
Sunday opens a particular position for the VivereMilano participants.
According to the manifesto, by deciding instead to stay in the city and
to care for it, the members of VivereMilano see themselves connecting
with a quintessentially Milanese identity and group of people. Ideally,
they identify with the Milanese who in the past decades would stay in
the city on weekends both to enjoy events such as performances at the
La Scala theatre and to meet each other in order to provide leadership
to Milan.

The forum's spontaneity – as a movement of people coming together
and talking to unknown others – and its passion to improve living conditions in the city have a utopian appeal and perhaps even constitute an
alternative model for participating in city affairs; however, these comments give a sense of a group of professionals claiming a role that they
see as being already rightfully theirs. Indeed, it is the combination of
these two aspects that makes this group so interesting for a discussion
on public space. Now that these middle-class 30- to 40-year-olds have
"woken up," they can almost effortlessly start talking and assuming a
position that has always been available for them. Politicians are listening to them, they have come to their meeting, and they might try to rally
them for their cause or party. Similarly, the newspaper has been encouraging them, and offering them space for expression and representation.
This is clearly not always the case for other meetings and groups, such

as the frequent immigrants' demonstrations or the actions of the Social Centres (see chapters 4 and 5).

Without implying that VivereMilano will necessarily be taken seriously by the municipality and the media, nor that they occupy a simple, uncontested position, nor even that the term "middle class" might suffice in characterizing this complex and heterogeneous group and its efforts, its participation and vision raises some interesting questions. What are the links between the underlying assumptions some of the forum's comments make about the city, public space, and its users, and the fairly privileged social standing of the VivereMilano participants? It is interesting in this respect to look more closely at the stone plaque (situated on the pillar under the street sign) under which the group assembled on 23 January (see Figure 19).

This sign, as well as other inscriptions nearby, reminds passers-by that the Galleria has been known, and presumably is still supposed to be, the "living room" of Milan. This is because it has always been considered the traditional meeting point of its residents and one of the most important places for social interaction. However, the imaginary of the living room – a location in the house that, in contrast to the kitchen, is traditionally reserved for polite conversation and display of social status (see Del Negro, 2004; Portelli, 1990) – reflects the unnamed/unmentioned specificity of the public envisioned as its primary user: mostly upper-class, white, Italian-born Milanese (men).

As Castellaneta writes, historically the centre was the domain of privileged classes, with workers and political activists constituting the "violations of this unwritten code" (1997: 99). He argues that lower-class people did use this space, but according to very specific roles and social connotations. Apart from activists and dissenters, lower classes were present in the centre either as workers, and identified as such, or as spectators of a display that saw higher classes as protagonists. Although in the last part of the twentieth century there has been, so to speak, a democratization of public space, and in contemporary Milan a wide variety of people use the central piazzas and streets, this does not mean that everyone is equally welcome in them. This feeling is reflected in the common complaint that the centre, and the Duomo Piazza in particular, "is not like it used to be," that it does not belong to "Milanese" people anymore – see also Dines (2002), for the discussion of a similar process in Naples – and especially that "immigrants have taken it over."

Gathering exactly, although perhaps not intentionally, under the inscription of the "living room of Milan," VivereMilano is in a strikingly

Figure 19. Stone plaque in the Galleria Vittorio Emanuele. Dedicated by the association Living Room of Milan to "the creator of this extraordinary architecture [the Galleria] in which is happily mirrored the genuine soul of the city," this plaque celebrates 100 years from the realization of this building, "where for a century Milan has lived all of its greatest hours, in joys and in tears." (Photograph by Enrica Sacconi, 2006).

easy position to resonate with the plaque's imaginary. Although the group did not refer directly to the sign and its expression, VivereMilano's use of and presence in public space does not challenge the plaque's reminder of the particular public that is supposed to be and speak there. On the contrary, it confirms it. Many of the participants' comments show that the movement seeks to recover public space for Milanese precisely by referring to traditional ideas of who is entitled to it. After all, the manifesto declares, the issue is not that these 30- to 40-year-olds have been marginalized from the city; it is just that they have never cared for what is legitimately, and as a matter of fact, theirs. By actively forgetting to remember that many people are excluded from the city, the forum of VivereMilano can work, paradoxically, as an invisible act of erasure.

What is interesting here is that it is exactly the enactment of public space as an ideal agora for the expression of city dwellers' voices that helps prevent other commentators from saying something, too.

Of Politics and Anti-politics[4]

These dynamics of entitlement and unequal participation emerged in interesting ways in the forum's talk – or rather, in its effort *not* to talk – about politics (meant as the conventional political and electoral system, but also encompassing wider political processes). One of the major themes running throughout the meeting of VivereMilano was the nature of the relationship between political representation and the movement's presence in urban space. The issue has been debated at considerable length on the webpage, too, as one of the cornerstones of the movement is its insistence *not* to be a political force. Indeed, the very presence of VivereMilano in public space that January morning was meant to symbolize a different approach to the problems of the city than the one offered by the conventional political system.

For many participants this seemed to be a crucial and necessary aspect of the association, but for others it seemed a source of bewilderment. One of the bystanders at the January meeting, for example, a local administrator of a nearby municipality, explained that he has always been involved in politics, and thus he came to the forum because he was curious to see who these people might be, who are the same age as himself, yet had always stayed out of political discussions.

As a bystander/participant in the meeting, I was deeply struck by the uneasiness that seemed to pervade the group whenever politics was mentioned, and I was fascinated by the sense that in the forum participants were talking about politics in spite of themselves. For in the very moment in which speakers stressed the independence of the movement from political structures, they necessarily brought to the fore the question of politics in and through public space (see Mitchell, 1996: 128). As one participant expressed it, there must be some relationship between a large group of people meeting and discussing openly in the street and the political process, but that link was not clear or defined. Using Don Mitchell's words: now that VivereMilano's participants inhabited this space "for representation" (1995: 115), who and what exactly did they seek to represent?

Analysing anti-government protests in Argentina in the 1990s, Guano (2002) suggests that public spaces help constitute groups and individuals as social and political subjects. People who could otherwise be seen as simply "private individual[s]" (306) and "politically irrelevant being[s]" (Arendt, 1958, quoted in ibid.) become and/or are seen as "public persona" (ibid.; see also Habermas, 1989). One reason why and how people and movements can claim this role is that public space works as a multiplicity of "performative arenas" (ibid.). Being in and using these locales necessarily involves an engagement with multiple audiences, with various layers of meanings and histories of spaces, and with different notions of citizenship and identity (see also Taylor, 1997).

Following Guano's insights, I argue that VivereMilano's evocation of the piazza as a locus for encounter and debate, and its participants' enmeshment with the very theatricality of public space, made it almost impossible to deny their position as both political actors (residents who are implicated in formal electoral politics as voters) and people who occupy public places in various ways. Paradoxically, however, because the link between claiming public space and doing politics was inevitably coming up, the VivereMilano meeting itself could be seen as an effort to keep questions of politics at bay: the forum itself sought to carve public space as a place where people could participate as residents instead of political subjects. Interestingly, this served to champion the concept of public space as residual instead of constitutive of political struggles and processes. Public space worked in this imagination as what could stand apart from politics, a place where "people" can meet and talk and from where "a-politics" (understood as the absence of political engagement or processes) itself could be constructed.

Two considerations are important in this respect. First, this focus on "a-politics" is especially interesting considering that Berlusconi's party and allies were ruling Milan by promoting an anti-political stance (Guano, 2008; Pasquino, 2007). Central to Berlusconi's rise to power was his insistence on running the country through managerial business practices and reducing the "fastidiousness" of "the presence and the demands" of the state (Pasquino, 2007: 51). As Gianfranco Pasquino notes, Berlusconi appealed to and reinforced an existing antipathy to politics, expressing "a remarkable disdain for the procedures, the regulations, the rules, and even the constitution" of the state (ibid.). Second, the forum's deep ambiguity in relation to the political – itself loosely defined – upheld a view of space as homogeneous, equally accessible to all, and constituted by a set of relations that are assumed to be already "settled" (Mitchell and Staeheli, 2005: 367).

This, to recall once more the plaque above, meant that the forum effectively effaced the specificity of the subject allowed to inhabit public space. The following invitation, posted in May 2006 by Cesare Fracca on the webpage, to gather for the second time in the same location as the January 2005 meeting is interesting in this respect. It reads, "Let us show again in Piazza del Duomo our love for this city, our will to change it and to render it more similar to us, to our children, and to our parents" (Fracca, 2006). I interpret this call to reflect the participants' desire to live in a city that offers a better quality of life to its inhabitants by being – according to a common Italian expression – *a misura d'uomo*: made with human "measures," that is, corresponding to people's needs and rhythms; literally, to live in a city that "fits" its inhabitants.

At the same time, this call is strikingly ethnocentric. The self who the city should resemble – the self reflected in *our* parents and *our* children – does not seem to include many unnamed and undefined others who also reside in Milan. The notion of public space VivereMilano suggests as fixed, neutral, homogeneous, and outside of political processes, in turn, normalizes the subject who is legitimately entitled to it: the "unproblematic," "standard" citizen, the parent or child of white, Italian-born, middle- and upper-class Milanese.

If public space always belongs more to some groups of people and to some "publics" than to others, ironically, in Milan this differential entitlement to the city manifests itself *both* in the way the middle and upper classes claim city spaces *and* in the way they ignore or renounce them – or perhaps, more precisely, in the ease with which they can move between these two positions. Better-off residents can marginalize themselves from public spaces (Bayat, 2012; Caldeira, 2000) – or at least say that they do – and choose not to engage with their diversity. This is because they have more resources than poorer residents to meet in and use spaces with more restricted access and/or to reach recreational destinations.

A white, Italian-born street vendor who has been working in the centre of Milan for decades, for example, told me:

> Here [in the Duomo Piazza] you do not see the bella gente [well-off people; lit., the beautiful people]. The bella gente meet in other places, they have cars and can meet wherever they want. (5 Nov. 2004)

Similarly, Alberta Andreotti notes that "the most marked form of segregation, [...] does not seem to involve the lower-income classes to the same extent as it does the upper-income classes, who segregate themselves in the more central and exclusive areas" (2006: 330; see also Bonomi, 2008).

What is interesting is that better-off residents' retreat from public spaces does not make urban locales any less their own. When they gather in piazzas, they are still able to make a compelling argument for their belonging there. Indeed, their ignoring public spaces as important locales for the negotiation of social and political citizenship can work to efface their very role and privileged position within them. As Gordon (1997) would say, the strategic absence of well-off Milanese in this context can be an important presence in the making of social reality.

Afterword: Another Shade of Orange

Because of the self-consciously "a-political" focus of the first meeting of VivereMilano, it was very striking to me when in 2011, six years after the forum I recounted above, I found myself in the Duomo Piazza in the middle of a gathering of another "orange" people. This time, however, this colour signified a completely different movement – namely, the political alliance supporting the candidacy of Giuliano Pisapia in the 2011 municipal elections. Party politics and the electoral system were in this instance at the heart of people's actions and motivations. Here I would like to briefly discuss this gathering and some of the debates sparked by the campaign and voting that preceded it in order to highlight some of the differences and similarities of these two "orange" moments. In particular, I am interested in how they brought "hope" – as an affect and a mode of engagement – into the piazza.

The 2011 campaign saw two main opponents vying for the position of mayor in Milan: the incumbent Letizia Moratti and Giuliano Pisapia, a lawyer with no prior political role (see Braghiroli, 2011, for a summary of the campaign). Municipal elections in Milan are usually followed by many citizens with nothing more than mild interest; this one, however, capitalized the attention, anxieties, and hopes of a large portion of the population, as many electors believed that the wider political life of Italy was connected to the results.

Moratti was the candidate favoured by the centre-right and by the Forza Italia Party of Berlusconi. It represented the continuation of the political status quo in the country. Many voters, disillusioned with Berlusconi – because of the economic recession, and ongoing reports of his involvement in prostitution circles and in shady dealings – were hoping that the demise of Moratti could serve as a sign to the Forza Italia that people wanted him gone. Moratti, moreover, had been criticized by the left and by most non-profit organizations for her regime of tight control.

Pisapia, instead, represented a new direction, and a change from the dominance of the centre-right. He was supported by a wide coalition, which straddled the centre and more oppositional groups in society. As Maurizio Mentana put it, "Pisapia is the perfect candidate, as he appears moderate to the moderates and radical to the radicals" (cited in Braghiroli, 2011: 149). This coalition, moreover, resulted in Pisapia being able to gain support from "young, semi-marginal voters" (ibid.) who might have not participated in the election at all. As one activist of a Social Centre commented to me a few days before the second election, Pisapia had become their best hope to see some change in the city:

> If even this does not work it will really be trouble; then we will not keep quiet any longer. (28 May 2011)

On the opposite front, the people supporting Moratti were fearful that Pisapia would bring about the further decline of security, legality, and moral standards in the city. Moratti had worked hard to curb crime, panhandling, and illegal squatter settlements. To many voters, Moratti represented Italian identity and cultural values, and she thus appealed to people who felt that new immigrants were posing a threat to the way neighbourhoods, families, and public spaces had been working for decades. Last, but not least, many Milanese preferred Moratti as she had achieved a number of important goals for the city, such as the awarding of the International Expo 2015 to Milan, the establishment of new green spaces, and environmental protection measures (Braghiroli, 2011).

For all these reasons, the election seemed to raise unprecedented passions. The very pragmatics of the vote, moreover, contributed to the heightened attention. The election was scheduled for 15 and 16 May with the understanding that if the results showed a close gap between the candidates,[5] the voters would be called for a second round, to happen two weeks later. When discussing the issue, then, people expressed hope and concern not only in relation to the end result, but also regarding the stages of the process. In other words, people did not just hope to win or lose. In this situation it became possible to lose yet feel victorious for having at least prolonged the incertitude and made the enemy feel that it had not been an easy gain. It was possible to hope to win twice, thereby making the result even more significant.

For people supporting Pisapia, this particular structure of voting, together with the high stakes at hand and the understanding that the centre-right was very hard to dislodge in Milan, contributed to the emergence of hope as a key affect and element of the campaign. The phrase *speriamo* (we are hoping), that I heard often expressed in relation to the elections, became a shorthand description of a range of options that could, each in its own way, signify a rupture with the existing political climate. The continuum of possible outcomes was widened by the fact that people were awaiting another important casting of ballots – a referendum that was to be held only a few weeks after the municipal elections. This referendum, too, was connected to the fate of Berlusconi, because its goal was to repeal laws that his government had passed earlier.

On 12 June, in fact, Italians would be voting on four referendum questions – one on indemnity laws for politicians, one on nuclear energy, and two on the privatization of water (see Chiaramonte and D'Alimonte, 2012). Adding fuel to the fire, the referendum itself was shrouded in a cloud of stormy debates, accusations, and uncertainties. For one, there was no consensus on whether the issue of nuclear energy would be on the list of questions at all. For another, it had been sixteen years (1995) since any referendum had been successful in changing laws, because of the difficulty of achieving the necessary quorum (the minimum numbers of voters – here, 50% plus 1 – whose participation in the voting process would make the referendum results valid). Thus, the people who supported the changes proposed by the referendum were concerned that not enough people would go to the voting stations. The statements of Berlusconi just prior to the referendum day that he would not go to cast a ballot further angered many of the voters. Many people who supported the referendum were upset that it had been scheduled so late in June – after the closing of schools for summer vacations, rather than at the same time as the municipal elections – and they believed that the government had delayed it on purpose to attract fewer voters.

Water, meanwhile, was taking on people's imaginaries and conversations. As people and organizations defined and discussed water as a public good, ideas and debates on what the latter could mean circulated in the city and were used to think of other issues. What else could "the public good" be? In the university, amid friends, and in everyday casual exchanges, people talked. Could public good be citizenship? Or education? And in what ways? As water opened the way for more issues to be debated, for people who were interested in a change to the political situation, hope itself took on a very particular role: it became an affect that could be evoked to open new spaces and occasions for questions and conversations.

Here I do not mean to point to hope solely as the wish for a specific result, or as a necessary ingredient of political rallying. Although we can say that every candidate hopes to win and every campaign reflects hopes for the future, "hope" often refers to an afore-seen plan of action that might come into place. The focus, in other words, is on the end product. Instead, here I am interested in hope as a process and an interruption, as what can create small yet lingering gaps in the order of things. I am thinking of hope as an embodied process of knowing, rethinking, and imagining, which does not focus on a set outcome, but rather introduces a discontinuity with the present and reorganizes the direction of our attention. This understanding echoes Miyazaki's (2004) description of hope in his ethnography of Fiji: a particular orientation towards the future, which leaves it open rather than already determined. This is akin to imagination as an "embodied practice of transcending" current places and circumstances (Salazar, 2011: 577), which shapes people's lives and choices by widening the possibilities they can envision (Appadurai, 1996; Schielke, 2013). In Milan, after the first round of voting, a pervasive sense of impending change was heightened by the fact that, at that time, the movement of the Indignatos in Spain was bringing thousands of people into the piazzas. For many residents of Milan who were interested in a different political direction in their city, signs of transformations elsewhere (including earlier popular movements in Egypt and the Middle East) helped them imagine that in Milan, too, long-established political alliances could be challenged from below.

This meaning of hope was brought home to me during many encounters, but perhaps one of the strongest instances was a casual conversation at one of the city parks. On the second day of voting during the first election, I approached a woman selling tickets at a children's play structure. I bought tickets through the glass window of the cabin she was sitting in, and I noticed that she was following with rapt attention a news report on the election. The results would not be announced until the evening. "Do you think Pisapia will win?" I asked her. "Of course we will win," she answered with fiery certainty,

> at least in the first round! Yes, we will go to the second round!! We cannot possibly always lose, sometimes we too can win! It cannot always be night!! (16 May 2011)

What struck me in this exchange was not only her defiant conviction, but also the sense that hope had become a permission to imagine that things could be different and that history could go contrary to what had been until now. Her certitude, the spark in her eyes, the excitement in her voice, caused me to wonder about what it takes, in this particular historical moment, to envisage change, and for this imagination to open spaces for discussions between people. Her sentence "it cannot always be night" that she repeated several times was an invitation to think that the future was no longer preordained. Her words were particularly striking because of her position: there she was, in a little bare room, in front of a children's play structure that was very often empty, a woman not young anymore, dressed simply, and looking tired. The kind of hope she communicated then seemed less tied to expected material changes than to a temporary opening of possibilities for thought within and between imaginary horizons. What I found particularly striking was that this feeling of hope, as a form of questioning and inviting, became a travelling mode of engagement. To say it with a metaphor, for those who were supporting Pisapia, hope in the city of Milan at that time was similar to a series of small fires that started to burn in different locales. It was showing up in unexpected locales, it shaped people's interactions, and it made room for intense conversations. Hope became a momentary sentiment that authorized inhabitants to speak about politics, society, and urban realities.

After Pisapia won the first round – as promised by the play structure ticket seller – passions intensified. People would spy on each other to try and guess how their interlocutor would vote. More orange flags appeared on buildings. A clothing store close to where I was staying dressed its whole window in orange, and advertised a 50% discount on any orange item of clothing it sold. Celebrations on one side paralleled the gloom on the other. The Moratti supporters felt that Pisapia did not have the political expertise to rule the city and to ensure the well-being of most of its residents and that, instead, he would let himself be guided by extremists. Posters against Pisapia showed up overnight, trying to appeal to sentiments against migrants, Muslims, and nomadic people residing in Milan. They displayed statements such as "with Pisapia Milan will become a gypsytown," and Milan will host "the largest mosque in Europe." It was not surprising that after such a heated campaign followed a very public and festive celebration when the centre-left finally won.

The Duomo Piazza became a key meeting place for supporters of Pisapia in several moments during the campaign, but the most well-attended and symbolically significant one was on the occasion of Pisapia's victory in the second round of the elections, on 30 May. That evening, thousands of people gathered in the piazza, wearing orange sweaters, scarves, hats, or bags, and displaying orange balloons, ribbons, and accessories. As I joined the celebrating crowd at five in the afternoon, the piazza was already full of people and hard to navigate. The crowd was heterogeneous: it comprised people of different ages, children, families, group of friends, and individuals. Although most of the people were white, many visible minority men and women were present, too. A large stage had been set up on one side of the piazza onto which speakers were taking turns in announcing the results (55% for Pisapia) and congratulating the city.

The other key location of the piazza was a large statue facing the Duomo. Dozens of people climbed the statue and were sitting on its tiers of steps, waving orange flags and balloons at the stage and the crowd around them. Under the base of the statue people wondered in amazement and awe at a stroller with newborn twins, each with an orange ribbon tied on its clothing. Some elderly women remarked that this gathering reminded them of the celebrations at the end of the Second World War. People were aware, moreover, that it was just the beginning of the party, as it was scheduled to continue into the night. The Duomo had become a destination for people travelling to the centre of Milan from many different parts of the city. Later that evening, for example, a rally started from the Buenos Aires neighbourhood where the Pisapia headquarters were located and journeyed through the centre and to the Duomo, echoing through the streets.

As I, too, sat on the steps of the statue witnessing electoral euphoria, I was struck by how this gathering was both similar and different from the previous "orange" meeting. In many respects, it embodied a radical change. If the forum of VivereMilano was an effort to build "a-politics" as a stance towards the city and the public sphere, Pisapia's victory party was a celebration of politics articulated and put into action through the electoral system. Indeed, May and June 2011 became memorable months in Milan because *so* much voting happened in the city. We could venture to say that in this time many residents rediscovered a sense of political passion. The fact that the referendum actually achieved a quorum further emphasized the political process and elections as instruments of transformation in the life of Milan and of Italy. Indeed, because the choice of mayor and the referendum raised such high stakes for the political destiny of the country, the time between the latter and the first municipal vote became "the thirty days that shook the North" (Anderlini and Goldstein, 2011: 585).

A shift towards the re-evaluation of the political process is what the Pisapia people wanted observers to see in the Duomo gathering. The movements against anti-politics and against Berlusconi were strongly interrelated and connected. The forum of the VivereMilano people has an important story to tell in this respect, as it shows the active, though contradictory, building of a-politics in the daily life of Milan. Indeed, only by realizing how subtle and pervasive a-politics and anti-politics had been, can we understand how different this movement felt and why it signified such a break.

In reflecting on these two different events, however, my goal here is to point out also their continuities and similarities. In particular, I am interested in the role of hope in both movements and in the role of the piazza in articulating these imaginaries. Hope as an effort to imagine change was a key ingredient of the two orange movements notwithstanding their contrasts. VivereMilano became an event in the city because it helped many regular people to envision Milan differently. It responded to a sense of confusion and frustration in many inhabitants, who felt that the problems of their city caused them to lead a "stupid" and contradictory lifestyle. Its sense of promise, at that first meeting, resulted from the idea that the Milan of tomorrow could be different from the one of today if enough people contributed to its vision and started to act together to directly change the current conditions. Indeed, the very enactment of the piazza as the agora where the citizens could voice their opinions was built upon this sense of possibility, even though it effectively marginalized other Milanese inhabitants and perspectives. Similarly, the Pisapia celebrations, and the campaign preceding them, expressed hope as a process of anticipating a future that was not yet charted and thus could bring in a rupture from the past. Interestingly, in the Duomo Piazza in 2005 a-politics was seen as making space for a different future in the same way that electoral politics could six years later.

More important for my purpose, it is hard to imagine these movements and their passions without the piazza; simply put, it seems that there is nowhere else where this can become tangible with such a force. Thinking of Miyazaki (2004), we could say that in both gatherings the indeterminacy of the open central place by the Duomo was crucial in embodying and communicating a sense of possibility. This, of course, does not mean that these or other movements and gatherings that use a piazza to search, propose, or celebrate alternatives are progressive, inclusive, or benefit most of the population. Rather, the relations between piazzas and hope is interesting because it both highlights the appeal of "public space" and the enthusiasm it can help generate among its participants and alert us to the very specific forms in which piazzas can take hold of the imagination as imaginations take hold of the piazza.

Indeed, what I want to emphasize here is that this very idea of hope can be read against the grain of the idealized function of public space as something that is inherent in the latter as a social construct, as something already there just waiting to be used. To say it differently, here I do not want to glorify the coming together of people in the piazza, to argue that one orange movement is better than the other, or to forget that repressive groups, too, use public space to embody passion and hopes. What interests me, rather, is how the two events I describe in this chapter are two very specific instances of the emerging of particular, hopeful imaginings in the piazza. What I argue, in other words, is for us to consider the very specific ways in which ideals are embodied, thought about, performed, and enacted in particular places and times. Even hope can mean different things in different contexts and be constituted through different processes and temporalities.

In turn, this raises helpful questions for researching and listening to other movements and gatherings: How do passions and desires get articulated and take a place in the city? How do they change and create new alliances? How are they spatialized? Whose dreams do they benefit? If public space is often imagined as an ideal, and as a locale for hope, the initial VivereMilano gathering helps us remember that the very way this imaginary is enacted and performed can serve very different agendas and social positions. Most importantly, it is not that VivereMilano used the ideal of the agora in a "wrong way." Rather, there is no ideal inherent in public space apart from what is contingently, contextually enacted and performed by people in specific places and times.

In a sense, the two events stand as correctives to each other. If the VivereMilano forum can help us understand why the 2011 Pisapia victory gathering felt like a crucial departure from the status quo, at the same time it cautions us as to its consequences. What shape will the political take in the next few years in Milan, and who will speak through it? What will happen to all those who hoped Moratti could win? Many people, including residents of the Milanese hinterland and inhabitants who are not Italian citizens, are not entitled to vote – what would their choices have been? More generally, who is included in the piazza gatherings, and how do such moments exclude others?

Chapter Four

Spatial Politics

Looking for public space in Milan, I was soon directed by many people to several urban activist headquarters that designated themselves as "self-managed public spaces" (see, e.g., www.leoncavallo.org). These Social Centres (*centri sociali*) are autonomous community sites that serve as venues for political, social, and countercultural grassroots activities. They are usually housed in (illegally) "occupied" buildings, and, while each centre forms a distinct entity, most of them are connected to others and constitute a wider alliance and action network in the city and in Italy.

The Social Centres' "marginal" yet "fruitful" (Mudu, 2004: 926) influence in the life of Milan can be seen in the large number of people who visit them (several thousands each month, according to Mudu) and in their ability to mobilize crowds for events and demonstrations in very short periods of time (see also Membretti, 2003). Most importantly for my purposes, since the 1970s these associations have been fostering an ongoing, though largely silenced discussion on urban spaces. Interestingly, by bringing attention to "invisible" urban locales such as dismissed areas (see also chapter 6), and by using them to construct public spaces, these organizations comment on and critique the politics of space in the Milanese territory.

Notwithstanding their important historical, cultural, and political role in Milan, Social Centres have been the constant target of police raids, public outcries, and municipal action plans. The media often describe them as breeding grounds for crime, drug use, and social unrest. Some city councils have tolerated their presence, but others have seen them as a threat to private property and urban security. According to Ottavia and Alberto, two activists from Naga – a free health clinic and an organization working for the rights of marginalized people – Social Centres constitute the quintessential enemies of the Moratti government. Alberto explained:

> The politics of this administration, which is from the [political] right since 1990, has taken two important phenomena against which to fight: the Social Centres [...] and nomad people's settlements. These two have been the only battle horses [i.e., the main expressions] of Milanese politics in the last twenty years [...] [Because these spark] fear towards those who are different: because the first are supposed to throw Molotovs, and the second ones are supposed to steal from people's houses, and abduct their children – just as the old fairy tales said. (14 June 2011)

Similarly to Roma/Sinti people and settlements, Social Centres and their sympathizers tend to be seen as disorderly subjects in disorderly spaces,[1] who trouble the look and feel of Milan as a middle- and upper-class domain. Social Centres were in the spotlight during the 2011 election, too. In the two weeks between the first and the second rounds of voting, the centre-right accused Pisapia of being an ally of squatters and their associations, thus creating a "communist" city and favouring illegality and anarchy. The constant tensions, for decades, surrounding Social Centres have indeed entrenched their position as resisting and oppositional subjects and strengthened their alliance with social movements ranging from feminism to peace activism, anti-racism work, and environmentalism.

In this chapter, I discuss interviews, events, conversations, and documents in order to follow this movement's insights with regard to public space and inequality. As I argue below, the Social Centres challenge a liberal conception of the agora as a realm of social debate – as the one, for example, enacted by VivereMilano. In particular, the Social Centres dispute the very possibility of public space as a realm of representation, political engagement, and sociality in the context of an unjust society. This is because streets and piazzas are always already colonized by the powerful. Indeed, the term "public space" commonly used in society is itself a way to conceal inequalities and exclusion.

From the perspective of these organizations, space becomes public only by serving as an access point for alternative understandings, grassroots politics, and non-hierarchical social relations. This line of thinking results in a specific theoretical orientation to space that focuses on the history of capitalism in Milan and on the emerging power relations of the post-industrial metropolis. Moreover, one of the aspects that interests me most is how the Social Centres' understanding, enactment, and construction of public space foreground memory and knowledge as key components of spatial politics.

In order to discuss these aspects, in this chapter I will listen to Alice and Giacomo, two of the many people who have participated in this movement over the years. Giacomo, a student, was among the founders of one of the newest Social Centres in Milan, called La Casa Loca (the Crazy House, in Spanish). I met him the first time I visited this Centre, and he introduced me to the building, its projects, and some of its dreams and strategies. Giacomo had recently moved to Milan from another city in Italy in order to attend university. Alice, a mother, grandmother, and activist, is a long-term resident of Milan. She is one of the Mothers of the Leoncavallo, an association of women who have been involved in the Centre Leoncavallo (the largest and most established Social Centre in Milan) for more than two decades. Today, most of the Mothers are in their seventies and eighties, but they are still very active in the Centre and in the wider movement, where they are very well respected.

Because Mothers like Alice have been part of most of the Leoncavallo's life and have, at the same time, witnessed the dramatic changes and struggles happening in Milan since the late 1960s, they represent, in certain respects, the historical memory of the Centre. Their work of remembering and continuing to tell what happened then, how it is connected to places and people, and why forces from the left and the right have always been part of the story is particularly important in the context of this movement. According to the Social Centres, in fact, it is only by "refusing to forget" that it is possible to uphold a vision of space as social and not just as a collection of places that can be changed and disposed of at will. As Alice used to tell me, however, forgetfulness is widespread in the city and is, indeed, a strategy for pushing to the margins the people and groups whose stories do not fit well with more official histories of Milan. For this reason, at the heart of the actions of the Social Centres is an effort to keep alive and present people and ideas that are rendered absent. This aspect of the movement emerged for me vividly in my first visit to the Leoncavallo. This is an excerpt from my fieldnotes for that day:

I arrive at the Leoncavallo, and I am immediately overwhelmed. It is its visual impact that strikes me first: the walls both outside and inside are covered with murals, graffiti, and writings, countering the grey and sombre look of the old building. Although it is a quiet morning at the centre, several people are busy in the café, the kitchen, and the many meeting rooms. All of them greet me in a friendly tone, and although they do not know me, they accept my presence as a matter of fact. When they meet one another, they call each other comrades.

After walking through one of the halls, I find myself in the internal court-yard. There are plants, tables and chairs, some children's toys, colourful flags. It is an inviting, relaxing place, drenched by sunlight. A large banner is draped on one of the walls, announcing: "Return the city to us! We want it back." I reach the café and there I meet Andrea, a man in his sixties, who tells me he works here as one of the caretakers. He explains that he is always at the Leoncavallo except when his other jobs keep him away. He leads me to the exhibition that I came to see, "The City That Will Come: Thirty Years of Move-ments in Milan," and leaves me alone in the vast hall. The walls are covered with large pieces of paper filled to the brim with dates, descriptions, drawings, and photographs. They narrate a history of Italy, of Milan, and some of the events unfolding in other countries. I remain spellbound both by the incredible vastness and richness of the chronology, and by the vivid sense of history and memory being enacted in this very particular space. It is a daunting presence, a weight I can almost feel in my body. I turn on the video camera that I brought with me and follow line after line, and image after image, its telling of battles, struggles, and movements that span the history of the city from the 1970s onward. However the recording of the material does not come close to its sheer size, or to the impact it has in this hall, which seems to be echoing with voices, stories, and lives.

After the exhibition, and having managed to read only a part of it, I go to the library. It is a small room, but filled to the brim with materials. I spend hours there, both reading documents and books, and talking to Flavia, the volunteer librarian. She tells me about her own journey to the Leoncavallo. She tells me at length about the situation of education in the country, the funding cuts, the repressive policies of the Milanese municipal administration. Her dedication and knowledge seem unbounded, and I listen to her for nearly two hours. She seems to know the whole history of Italy. (9 Dec. 2004)

In keeping with the Social Centres' insights about the political role of memory and forgetting, part of this chapter is intentionally histori-cal. It gives attention to chronicles: particularly situated recollections and analyses aimed at reformulating the histories of places, people, and events in the city. The exposition "The City That Will Come" (see Figure 20), organized and curated by the Leoncavallo in 2004–05, is a

Figure 20. Two images from the exposition "The City That Will Come: Thirty Years in Milan, the Metropolis of Movements," hosted by the Leoncavallo, that reconstructed the history of Milan from the 1970s to today from the perspective of Social Centres. Through the juxtaposition of photographs, dates, and analysis, this chronicle links spaces with events, people, political developments, and social conditions in Milan. In the photograph at the *bottom*, e.g., the exposition recounts the history of the Piazza Fontana bombing, with images, artwork, and a description. (Dec. 2004).

telling example. As its title indicates, historiographies are not only a way to map the past and the present, but also a way to try and shape one's future. This is because, as particular forms of remembering and retelling, they serve as interventions into long-standing debates and discourses (Roseman, 1996; Portelli, 1990). Specifically, the chronicles I discuss here are a way to theorize the relations between city subjects and the urban territory.

The commentaries of Alice and of Primo Moroni, a Milanese counter-cultural writer and scholar, which I discuss below, are other instances of chronicles. One of the things that struck me in my conversations with Alice is how her description of the Leoncavallo and its role as it is today could not be divorced from a detailed recounting of its beginnings, its many evacuations, and the phases of its existence. In other words, the very telling of the history of the organization was, for Alice, a crucial way of participating in it. Moroni, a renowned left intellectual and activist, was especially involved with social movements in Milan, and upon his death he left an impressive archive of writings. In this chapter, I draw extensively from one of his studies, because his work represents, in many respects, an alternative historiography of space. By echoing their telling of history, this chapter seeks to foreground the processes of knowledge production that shape and are shaped by spaces, and the way in which they intersect with my own ethnographic learning.

Remembering, in turn, motivates actions and shapes strategies. In the second part of this chapter I discuss some of the initiatives and dreams of Casa Loca, including its critique of neoliberalism and of precarious working conditions in Milan. Although – and, indeed, precisely because – these experimental places constitute public spaces that are often not counted as such (hence, the predominant view on the part of the municipality that Social Centres are usurpers of private property rather than providers of accessible social spaces and services), they offer very valuable perspectives for thinking about what "public space" could mean and how it could be constituted by different actors in different contexts.

Remembering: Alternative Histories of Space

Andrea Membretti (2003) has said that there are more than a hundred Social Centres in Italy, and Mudu (2004) estimates that in the years between 1985 and 2003, more than 200 Social Centres were created in almost all regions of the country (2004: 928).[2] The Social Centre movement is a very heterogeneous, "segmented," and "policephalous" body

(Membretti, 2003: 79). Centres are organized differently from each other, focus on a variety of goals and actions, and relate in different ways to the particular communities in which they are located (see Membretti, 2003: 72ff; Moroni, 1996).

Although Social Centres escape easy definitions and common descriptions,[3] they nonetheless constitute a social, cultural, and political movement. According to Membretti (2003), some of its central aspects are an important youth component; the illegal occupation of buildings to transform them into liberated, self-managed community spaces; an internal organizing structure usually based on non-hierarchical, egalitarian relations; and the creation of alternative forms of sociality, of work, and of cultural and material production (72ff).

Squatting – the illegal occupation of buildings to transform them into "liberated," self-managed community spaces – is often a defining characteristic of these associations, and it links them to wider movements active in reclaiming abandoned or underutilized buildings and sites in other European contexts as well. Squatting also is a strategy employed by poor inhabitants of Milan to secure a place where they can live; however, here squatting is a tactic explicitly used to protest urban policies and processes. To say it differently, even if Social Centres often provide accommodations for people who have nowhere else to go, they are first and foremost community centres that work as a resource for all interested city dwellers.

The Social Centre movement resembles a flexible, ever-changing net – in the words of Mudu, a "multi-centred non-hierarchical affiliation network" (2004: 927). The importance of participants' personal networks, alliances, and friendships is evidenced by one of Alice's joking comments during one of our conversations, that even when she and her family go away from Milan to spend their holidays, they find themselves visiting other Social Centres and/or friends involved in them.

The Social Centres in their current form were started in the 1970s. Strongly associated with working-class struggles, they used, by and large, communist idioms and ideals and were informed by the long history of conflicts that since the Second World War had divided political allegiances into fascism and the resistance to it.[4] The late 1960s and 1970s were an especially troubled time for Italian politics and society, with an escalation of conflicts at all levels. Social Centres were very active in these struggles and debates. The 1980s were a time of underground activities for most of them; in Milan, in particular, heroin became a primary concern for many youth in the city and for many activists from the Social Centres who organized campaigns against drug trafficking while seeing many of their friends killed by it. The 1990s, however, brought a re-emergence of Social Centres with a renewed opening towards the wider public. New groups and services were created, showing a shift towards new areas of research and action, such as immigrants' rights, precarious working conditions in Milan and in Italy, free trade and neocolonialism, and the patenting and commercialization of knowledge.

These changes have been shaping the relationship between Social Centres and the urban terrain. The oldest Centres were at first conceived as a part of the axis factory–home–community centre. They provided services to working families that ranged from health care to recreation. As the repression of organized labour increased, and the latter started to lose ground in the city, Social Centres could be seen as replacing the lost public spaces: the piazzas and streets where working-class interests had been debated, demonstrated, and fought for (Mudu, 2004). The issue of space became particularly central as Milan was closing its factories and its economy was changing in favour of the tertiary sector. Ironically, in fact, the diffusion and "multiplication of 'productive spaces' in the metropolis did not at all mean that there has been a multiplication of public spaces where it is possible to exercise political action" (Vecchi, 1994: 6). In this context, Social Centres became instruments to invent and construct alternative public spaces that were/are explicitly political spaces.

Among the Social Centres, the Leoncavallo in Milan is the oldest, best known, and largest in Italy. Started in 1975 at another location, the Leoncavallo occupies a large building, a disused printing factory, in Via Watteau, a peripheral area in the northwest part of the city. It includes two large halls that can be used for concerts and expositions, a kitchen, a cafeteria, a bar, a café, offices, a bookstore/library, and a green courtyard that can be used in summer. The many organizations, committees, and groups that operate within the Centre are involved in various areas ranging from cultural and media productions to the liberalization of drugs, sustainable development, and the rights of migrants. As well as initiating and carrying out social and political campaigns, organizing concerts and performances, facilitating public fora, and holding regular discussion groups, the Leoncavallo operates a legal counselling service and an Italian language class, a radio, and activities for children. The fact that up to a hundred thousand people a year visit the Centre is a good indication of its vitality and the range of services it offers (Membretti, 2003 and 2007).

The current Leoncavallo building was occupied in 1989, after squatting in other premises, and subsequent evictions by the police. Although in recent years, the Leoncavallo has "regularized" its position in relation to the municipality – see Membretti (2007) for an analysis of this development – causing important rifts in the community, it is still in continuous danger of eviction. The words of Alice vividly describe its turbulent history and the struggle for its very existence, as well as her own involvement with the Centre:

> I was working. At a certain point I heard at the radio that they had killed two youths, Fausto and Iaio. I immediately stood still. Two youths that visited the Social Centre Leoncavallo. Considering that at that time my daughter was already 14 years old – 13 to 14 years old – and from time to time, with friends, when there were concerts, she went there, I said: "Oh my god, what happened?" And then they said the day of the funerals [...] I went, I telephoned the boss [at work] and said: "Tomorrow I will not be there, because I will go to the funerals" [...] And that one, he said: "But Mrs, they were two criminals." And I said: "Let's keep calm; that is to be seen. I will start by going" [...]
>
> I went, and I was surprised by how many people were at the funeral of two youths of a Social Centre that was not even known yet. It was full full full. The people could not fit in the church anymore. The people who did not fit in the church were occupying the whole Street Casoretto [...] I was shocked, and I said, "Here I have to understand what it is that is not right" [...] So I went to see this Centre that was supposed to be a cove of criminals, a cove for deranged youth, etc. and I realized that it was an error. I had the fortune to meet the Mothers [...] so they told me the story.
>
> This Social Centre has been occupied because it was a dismissed area that did not serve anyone. In practice, some youths used to go, they drank a beer, and so on. So the Mothers said, "Come on, here we don't have anything. For the youth have practically nothing; let's help them clean [the place] so that they can stay there." So there they let them stay, they [the authorities] did not move. When the fact happened, of the two youths that were killed, the authority moved.
>
> Consider that in the same moment that we went to the funerals of Fausto and Iaio, in the house of the two youths, in the same moment that the funerals were going on, the police went to the house of one of the youths and they ransacked the whole house [...] to see where these youth were hiding the weapons that they supposedly had in their house ... they did not find anything ...]

Then every fifteen/twenty days we would go to the court [...] nobody knew anything. At a certain point it was discovered who it was [who killed them]. But we could not have the names of these people [...] They knew who it was, but one did not know, that is, officially, one did not know. We had our own information sources, and they told us that they could not do it [to tell the names] because the order had come from a very well-known public office [...] From that moment, we kept the court case open, but one did not know anything. After eighteen years, they wanted to archive the case. So they archived it, and we reopened it [...] It went on for very long time [...] The association Mothers of the Leoncavallo was formed for this reason. To have more access to public offices. (18 Jan. 2005)

According to Alice, the history of the Leoncavallo cannot be disasso-ciated from that of Fausto and Iaio – two youths killed on 18 March 1978 on the street while they were returning to their home. Although the assailants remain unknown, many people believe that they were mur-dered because of their research on and activities against drug traffick-ing in the area. Groups from the extreme right are suspected of having been involved, although until now, there has been no resolution to the puzzle or a publicly accepted version of the events. Many people of the left and the community of Social Centres see the murder of Fausto and Iaio, more generally, as one episode in a long and complicated history of violence that has earned this period the name of the "Bullet Years" (*Anni di Piombo*, or "Years of Lead") of Milan (see Ginsborg, 2003). This included a number of shady dealings against the left by right-wing forces (see ibid.). The Mothers of the Leoncavallo, by keeping alive the memory – and court case – of Fausto and Iaio, have been refusing to forget this time (and its continuities with the present), the struggles that have marked the city and its spaces, and the need to find answers to their questions. At the same time, keeping the court case open is an action meant to defend the Leoncavallo's right to exist and to propose a different set of ideals and values from the ones promoted by the powers that rule. Alice explained:

At a certain point came the fearful eviction notice of September[5] [...] Every time that they destroyed ... every time that they were supposed to destroy it [...] tomorrow morning ... this afternoon ... there will be the forceful evic-tion of the Leoncavallo ... we always succeeded to know it a few hours in advance, sometimes a few days in advance. At that point one called the associations that are in all Italy, and they would come to give us a hand: [...]

thousands of people [would come] [...] we would make a rally that went to the centre [of the city] [...] and every time, because there were so many people, they couldn't do it [the eviction] [...] At that point we had organized ourselves: we need people who stand on guard [...] They did not leave the Leoncavallo, neither at night nor during the day [...] On a tragic morning they call: come at once to the Leoncavallo [...] my husband accompanied me, and when we went there, it was full full of police [...] They entered by force, the police, and they destroyed everything, everything: the computers, the kitchen, the desks, the paintings [...]

After this famous eviction that they succeeded in doing [...] they gave us a little building, [...] in Via Mecenate [...] We went there, [...] Then one day, we were just coming back from holidays, the 15th of August, and I hear the TV that says: "In this moment there is the eviction of the Leoncavallo." They had taken advantage of the fact that we were all away [...] I jump down from my chair: "[her husband], take me there!!" [...] We did a big rally, [...] then we all camped in a park – fortunately it was summer – in a park that was close to Via Mecenate [...] and we stayed there night and day, we took turns [being on guard]. We [...] made huge pots of pasta and we brought them [to the tents] at noon, with bread, ham, some things ... Then [...] we went to Martesana, another fifteen days at Martesana. There, too, a never-ending solidarity [from people and groups]. Then at a certain point they phone and they tell me: "Come [...] to Via Watteau." We went there and there were lots of people; the youths had entered [a disused building, and] put out the flag [...] Because we were so many, the police [...] did not attack. So that evening we organized a big concert [...] free for all. (18 Jan. 2005)

One of the aspects that most struck me from Alice's narration is her and other participants' active engagement with the issue of space in the city. Our very first encounter is a vivid example of this. On a cold and grey weekday morning, I went to a residential building in a peripheral area of Milan to meet another one of my interviewees, an elderly woman who lived there with her husband. Unfortunately, that morning my interviewee had forgotten our appointment, and so I found myself wandering in the courtyard and outdoor passages instead. This is where I met Alice, who approached me with a friendly smile to ask me if I needed some help. Thanking her for her offer, I explained that I had come there to meet some people in order to talk with them about public space. "Public space?" exclaimed Alice enthusiastically,

I can tell you about public space! Public space, we occupy it!!! (18 Jan. 2005)

As Alice made clear, public space cannot be taken for granted, but rather it has to be made and reclaimed. Public space is the result of concerted efforts, militancy, and political action. Squatting, as a model for its construction, has to be seen in this light. Pierpaolo Mudu writes, "Squatting is an essential component of the strategic mix of these Social Centres [...] because it is a way of obtaining what has been denied" (2004: 922). According to Alice, the city denies its residents, and working-class youth in particular, places of sociality that are accessible, affordable, and foster equality between people.

Starting from an affordable recreational space, the Centre seeks to build an alternative model of sociality that includes the recognition of diversity and the extension of citizenship rights to all residents of the city (Membretti, 2007). With these goals in mind, occupying spaces is both a way to draw attention to issues of urban planning and inequality – and more generally to the social, cultural, and economic construction of space – and a way to create actual "public spaces" understood as sites of empowerment.[6]

The work of Primo Moroni highlights this double engagement with the urban territory: the ongoing struggle to construct oppositional spaces within it and the work of understanding, representing, and criticizing Milan's complex and dynamic spatial politics. Mapping the creation and distribution of several generations of "oppositional" centres (including but not limited to Social Centres) in the Milanese area, he points to the deep linkages between the creation of oppositional associations, the industrial and economic development of the city, and the "social history of [...] urban spaces" (1996: 167).

Moroni argues that because of the circular and spiralling shape of Milan, its residents experience and move within the city along triangles "whose peaks insinuate themselves into the historic centre while the bases [...] widen within the suburbs" (164). He describes two of these triangles, one located in the south and one in the north. The way in which Moroni differentiates between the two triangles is especially interesting, because it is an effort to trace a history of space conceived as an important component of social action.

According to Moroni, although both triangles include important working-class neighbourhoods and represent the route of approach of the lower classes to the material, institutional, and symbolic centre of the city, the South Triangle is closer to the centre than the North Triangle is because of the uneven growth of Milan. The former, in fact, has its peak in the heart of the historic centre, very close to the neighbourhood of Ticinese-Genova.

The North Triangle, on the contrary, has its peak located much further from the centre, close to Piazzale Loreto and Corso Buenos Aires. From this location, the route to the historic centre leads through avenues that are "empty and unwelcoming," and through San Babila, which is "one of the most elitist and inimical [piazzas] of the whole metropolis" (Moroni, 1996: 165). Because of this situation, "the big working-class neighbourhoods of the north and northeast [...] are much more distant from the historic centre [of the city] than the ones of the southern areas of Milan [...] But the first ones are not only topographically more distant [from the centre]. They are also located in an urban situation that includes more 'obstacles,' more 'foe' territories between the inhabitants of these places [the northern neighbourhoods] and the use of the historic centre, living 'soul,' centre of power, and place of innovation for the life of the city" (164).

This difference between northern and southern Milan, writes Moroni, is even more marked if we consider that the South Triangle includes a route of approach that is decisively "friendly" (167), because it is lined with stores, clubs, and inexpensive restaurants affordable to working-class people. This is strengthened by the social characteristics and particular history of the Ticinese-Genova neighbourhood,[7] which constitutes the "peak and heart of the South Triangle" (ibid.). For these reasons, writes Moroni, this area became "an exemplary urban frontier zone between centre and periphery" and between different classes, and "in the first years of the seventies the European neighbourhood with the highest concentration of extra-parliamentary political bases" (ibid.), grassroots political organizations of the left led by young activists and workers.

Another key observation of Moroni is that the symbolic and political meaning of the historic centre, to which the triangles point, has been changing through time, thereby sparking different forms of action by oppositional groups and different relationships between the latter and the territory of the city. An important component of these developments were changes in the relation between labour and capital in the region. In the 1960s and early 1970s, the industrial working class had more of a definite identity and bargaining power; however, in the 1970s a crackdown on organized labour caused a crumbling of what Moroni calls "extra-parliamentary political bases" (1996: 167), activist associations allied with workers' struggles.

The political and economic changes of the country caused these organizations to become disillusioned with the very possibility of acquiring institutionalized power,[8] embodied and symbolized in the city centre. As factories closed down, and as industrial production became smaller, more decentralized, more disseminated, and less labour intensive than before, new groups and action centres – called *"circoli"* by Moroni – were established in the suburbs. These associations, by younger activists who Moroni describes as "inexorably destined to the informal economy" (169), were guided by a desire to realize new social spaces that could represent them – spaces closer to the suburban neighbourhood they grew up in and that ignored the city centre as something distant and "irrelevant" (170).

These groups, suggests Moroni, have thus a different relationship to the city centre than the older generation of activists. They use the triangles to go to the centre to engage in political actions rife with irony and creativity in public streets and piazzas and to demand access to goods and services that they are economically barred from. The aim of these groups, however, is "no longer one of competing with political institutions" associated with the urban centre, but rather to obtain other "spaces, and territories that they can control themselves" (172).

I want to open a parenthesis here to point to a particular event of these years. One of the most dramatic and well-known incursions of the left movement into the historic centre of the city was directed exactly at those cultured Milanese evoked by VivereMilano: on the evening of 7 December 1968, a group of students led by Mario Capanna, and part of the wider worker/student movement of the time (see Ginsborg, 2003), attacked the people who were about to attend the prestigious First of the Season at the La Scala theatre, by throwing eggs and vegetables onto their very elegant attires.

This incident shows that "Scala-goers" can evoke widely different imaginaries for different groups and at different times. For the students of that time, they represented the Milanese bourgeoisie who was denying them full participation in the city and society; however, for some participants of VivereMilano, the Scala-goers can work as an example of residents who were interested and engaged in the city and provided it leadership. If Scala-goers can be the object of starkly divergent recollections of the history of the city, so can its spaces. We might want to remember here that the Galleria and its plaque are at the very heart of the centre of Milan. VivereMilano sees this as a simple and a-political space where "regular" citizens can gather to talk about the city; however, the intricate and changing significance of the historic centre for oppositional groups complicates this perspective and shows, once more, the historical and social specificity of VivereMilano's project. Like the one of the oppositional groups, VivereMilano's use of the space of the Galleria relies on particular interpretations and recollections of the history of Milan, and it is connected to categories like class, age, gender, race (and more).

The Social Centres were born of these particular "urban geometries" (1996: 173), described by Moroni. Started by the older activist generation that grew disillusioned with the strategy of competing with the institutional centre, and larger than the *circoli*, which proved to be rather ephemeral, they are similar to the latter in that they were formed in working-class areas and in dialogue with the social and political dynamics of each of their neighbourhoods. Some of the most important characteristics that the Social Centres share with the circoli are, according to Moroni, the self-government of their structures, "the necessity of being rooted in the territory," and the shift of focus "from the problem of time to one of space" (178). As Marco Grispigni explains, in the context of industrial work the goal was to free one's time in order to be able to use the available spaces: in a deindustrializing society, where many youth cannot find work, the issue is rather to obtain spaces, as well as income, "in order to enjoy the great quantity of time free from work" (1994: 28).

As many suburbs were restructured and many of their inhabitants relocated, the activists' relationship with the city became even more complicated. The changes were making impossible a "proud belonging to the periphery [and] to the spaces of memory of the *circoli*" (Moroni, 1996: 179), as the working-class areas of Milan were slowly being transformed by gentrification and a shift to a service economy. The geographical divisions between workers and ruling classes were becoming more and more muddled and diffused. As a consequence, the Social Centres increasingly saw *all* of the different areas of the city (not only the ones where they were located) – from the most central and elitist to the most abandoned and underprivileged – as mediums and results of social and political struggles. In this way, the historic centre of Milan became once again important for political action, although in a very different way from the alternative political centres of the late 1960s.

I discuss Moroni's fascinating account at length here because it crystallizes a very particular way of thinking about space. Moroni's cartographic history shows that the very existence and constitution of Social Centres over time is itself a reinterpretation of urban space. It is a practice aimed at rendering visible what is usually erased: the productive

relations embedded in and shaped through space and the class forma-
tions that go along with it. His analysis recalls Henri Lefebvre, who
argues that "space is permeated with social relations; it is not only
supported by social relations but it is also producing and produced by
social relations" (1991: 286). Basing this perspective on a Marxist analysis of production, Lefebvre insists that space is not just an innocent
stage onto which society unfolds, but is rather an important component
of social action.

Most importantly, Moroni's urban geohistory shows that Social
Centres are public spaces because they are locations from where the
city can be apprehended, interpreted, and acted upon. Public space is
here understood as a material and conceptual site – itself constructed
in struggle – from where action and reflection are made possible. Processes of knowledge creation and the writing of an alternative geohistory like the one of Moroni are essential in this respect. By going to
the Leoncavallo and engaging in long conversations with Flavia, the
activist who was volunteering in its library, is how I found Moroni's
work in the first place. Both philosophically and in a very pragmatic
way, Social Centres see and call themselves public spaces because they
are points of departure for accessing critical knowledge and developing
oppositional perspectives.

The "culture wars" that were taking place in the winter of 2009 constitute an interesting anecdote in this context. When I arrived in Milan in
January, I found a pervasive discussion on the role of culture in Milan
and in society at large. One of the things that accompanied the debate
was a series of large posters displayed all around the city. Each poster
consisted of a black plain background with a question written in white
characters. The interrogations included: Is culture politics? Is culture
necessary? Looking for Gramsci in Milan. What is culture? These provocative posters, which did not reveal to the audience who had produced
them and why, were part of a public art installation by Alfredo Jaar. At
the same time, the newspapers started a running debate on the position
and role of culture in the city. People deemed important for this topic –
local politicians, municipal councillors, professors, and artists – were
interviewed, and many of them argued that the city needed more funding for the arts, that the museums should open until later at night, and
other related issues.

What became immediately apparent is that the commentaries in the
newspapers represented an interesting shift in the very meaning of
culture. The posters of Jaar, with their reference to Gramsci, and their

insistent interrogations, conveyed a more open idea of culture, one that could include counterculture, as well as all kinds of creative and critical questioning and the exchange of information and ideas that happens in everyday life. Most of the commentary in the newspapers, in contrast, privileged an official understanding of culture, as the outcome of education and expressed in institutional art and literature. To criticize this situation, soon a variety of groups and association formed their own culture debate circles, questioning who should decide what is culture and in which contexts.

The debates, however, became particularly heated when in the middle of all this, on 22 January, the municipality closed down the Social Centre Cox 18, and its adjacent library, which hosted the archive of Primo Moroni's writings, confiscating all of its books and materials. Although eventually the collection was returned to the Centre, for long weeks its fate remained uncertain, and activists feared that the municipality would simply destroy it. The bitter irony of promoting culture and debates about it, while at the same censoring one of the most important collections of countercultural writings in the city, is telling evidence of the continuing tension between the mainstream and counterculture and between Social Centres and official definitions of public space, education, and knowledge.

Alternative libraries and archives are, not surprisingly, a crucial part of the Social Centres I visited. When I journeyed to the Temporary Stecca in 2011 (see the conclusion at the end of this book), for example, I noticed that a very significant part of the Centre was dedicated to a carefully organized collection of writings, documents, and publications on the Garibaldi Repubblica area, the urban renewal projects, and issues of gentrification. The "culture wars," moreover, bring us to rethink critically where "culture" is made, what is a common good, and what roles urban space can play in these engagements.

Acting: Social Centres as Public Spaces

As Moroni and Alice argued above, oppositional activists and commentators engage with the urban territory in a double way. By critically examining urban spaces, and especially the ways in which different kinds of locations sustain, intertwine, and/or interrupt each other, they bring into view a social history and "memory" of space that mainstream discourse and city officials tend to erase. Moreover, participants of Social Centres intervene in the production of public space itself.

Social Centres see space as a medium and result of social action. As Andrea Membretti explains with regard to the Leoncavallo, a Social Centre is a political subject by being at the same time "a territorialized actor" (2003: 70). In this way of thinking, urban spaces are not necessarily public by being formally and administratively so, or by not being privately owned. For one, according to Alice's words, contemporary public spaces are often sites that are indeed private but have been occupied and *made* public. For another, public space is in this context better understood as a living network of places – some public, some private – realized by particularly situated people. As in Moroni's account, this territorial web acts as a public space by allowing the expression of particular subjectivities, historical and theoretical perspectives, and social relations.

This focus on liberated spaces informs the recent creation and innovative countercultural projects of Casa Loca in the northern neighbourhood of Bicocca. If the Leoncavallo is the oldest Milanese Social Centre and embodies a very long and complicated history, Casa Loca is one of the newest Social Centres in Milan. Similarly to my conversations with Alice, one of the very first things that Giacomo, one of the founders of Casa Loca, told me when describing the Centre is the history of its very beginning – grounded in the act of "occupying" an empty building:

> In 2003 [...] we were in Chiapas [...] [It] was this student trip [...] We had a fantastic group in Chiapas. We return and we say: 'Why don't we start occupying a space in Milan?' (11 March 2005).

After occupying the building, the group organized the space as affordable student housing, responding to the lack of accommodations for non-Milanese students in town. Although at the beginning, the squatting simply provided a place to stay for the initial group of occupiers, after a few months, the latter managed to "launch" it as a "public space: [...] not a house for us, but a service [for others]" (Giacomo, 11 March 2005).

As a result, since 2004, Casa Loca has been operating as a self-managed, low-cost residence for persons attending the nearby University of Milan-Bicocca. Often these resident students become part of the organizing collective of Casa Loca during their stay and participate actively in its projects. The services of Casa Loca include an inexpensive cafeteria, catering mostly to the students in the area. The Centre is also linked with immigrants' rights associations, and it hosts a free Italian language class for migrants.

A key action in its history has been its support of another occupation, in 2004. Through the organization Action Milano, some of the members of Casa Loca helped twenty-four Latin-American-Italian families take possession of the so-called Plastic Houses (Case di Plastica). The latter, a building owned by Aler, the municipal body responsible for social housing, and constructed with experimental materials and innovative floor plans in 1970, had been left vacant for ten years, waiting for needed, but never realized renovations. Following the occupation, the families that started to live there organized and carried out some of the renovating themselves. The activists of Casa Loca became allies of the families and helped them to raise support for their cause and to avoid being evacuated.

Because the Plastic Houses are architecturally a pioneer work, and because they were occupied by immigrants, who, as a category, are one of the groups that experience the most difficulties in procuring accommodations, for the activists involved in it, this action drew attention to several issues at once. It spoke of migrants' rights and issues of poverty and marginalization. It revealed the incredible waste of resources in Milan, in which buildings stand vacant and lands unused, in spite of the lack of affordable housing for many of its residents – and irrespective of the fact that accommodations problems are one of the most significant factors in the impoverishment of lower-income residents (Bonomi, 2008: Zajczyk and Cavalca, 2006).

The activists of Action Milano and Casa Loca saw in the occupation of the Plastic Houses a way to reappropriate creativity and specialized knowledge, such as architecture, to profit the wider community rather than a small privileged sector of the population and to engage critically with the city and its territory. This was part of a wider goal of involving students and researchers from the nearby university in collaborative projects aimed at documenting and reflecting on the changes of this and other neighbourhoods.[9]

Another focus of Casa Loca, and of many other Social Centres in Milan, is concern with what Giacomo calls the current "precariousness of existence." As Molé (2010) discusses, this condition does not only refer to the increasingly unregulated, temporary, and flexible working conditions of many people in Italy. Precariousness is associated with a wider anxiety about the advent of neoliberalism in the country, where only a generation or two earlier, permanent employment represented the cornerstone of citizenship and signalled inclusion in the national community. According to Molé, as Italy makes the shift to neoliberal governance, the notion and experience of precariousness become a way to identify one's social fragility: "Avowing oneself as a 'precariat' signals a classed political subjectivity: a worker at the mercy of risk, marginality, anxiety, even paranoia" (2010: 39). Because of this, precariousness has emerged as a political discourse and social movement in Italy, "a publicly circulating term to critique neoliberal labour regimes" (38).

Casa Loca has been organizing campaigns on this issue, and many of its activists including Giacomo have been involved with the apparitions of San Precario. The latter, a rhetorical, symbolic, and highly carnivalesque figure, is the "patron saint of precarious, casualised, sessional, intermittent, temporary, flexible, project, freelance and fractional workers" (Vanni and Tarì, 2005, no page number). Vanni and Tarì describe its first appearance as follows:

> Shoppers don't quite understand why there is a procession at the deli counter of the supermarket. On closer inspection the statue of the saint is a bit odd. First, the saint is dressed as a supermarket worker. Second, it has too many arms. Third, it holds a telephone, newspapers with job ads, and McDonald's chips. The statue is carried on sticks by a group of young people, and a priest, a friar, and a nun are with them. There is even a cardinal. They distribute saint cards: San Precario is the name of this saint. Most people haven't heard of him. But then the young people say a miracle has happened and there is a 20% discount on shopping today. And with prices going up every month – prices have doubled since the euro was introduced – and the superannuation money being always the same, and the grandkids who cannot find a job for more than three months even if they went to university and studied law. (No page number; describing a scene in a Milanese supermarket on 29 Feb. 2004)

As Giacomo recounted, the figure of San Precario is meant to protest increasing deregulation of working conditions in Italy. It is, moreover, connected to wider critiques against contemporary work and sociality, global politics, and division of labour. For these reasons, explained Giacomo, San Precario has become a key "icon" for the movement of the Social Centres as the saint "embodies the analysis" that these organizations have been developing for several years (11 March 2005). Although San Precario was not born in Casa Loca, but is the creation of another group, the Chainworkers, Casa Loca has been one of the places in which he appears and has provided the elusive saint followers and support.

The work of Casa Loca, such as its focus on the reappropriation of knowledge and its involvement with San Precario, is emblematic of the shift of Milan from an industrial centre to a post-industrial metropolis. This is strongly reflected by the very location of Casa Loca – geographically and symbolically – at the mid-point between an old working-class neighbourhood and the new developments of Milan-Bicocca, which include the university as well as expensive new residences. Giacomo explained:

> On this neighbourhood, on Casa Loca in itself, on the possibilities of this space, and the ideas that sustain it, I think that by looking to the other side of this street [the Sarca Avenue, where Casa Loca is located] one understands the big significance that this space [Casa Loca] has for Milan [...] [This is] one of the neighbourhoods on which people are promoting big speculations, new funding, development of chains of shopping malls but not only, the use of knowledge produced in the university by multinational companies [...] [This] is all a context in which [our] [...] presence is very interesting. And for us to work in this neighbourhood is very beautiful [...] also [with regard to its] human fabric, [there is an older population] that still resists [the developments].
>
> Already the Sarca Avenue [where Casa Loca is located] divides in half the Bicocca [neighbourhood]. On that side there is the super exclusive part where one apartment costs, I don't know, four–five thousand euros per square metre, and on this side there is the old part, the social fabric of factory workers who lived here since the 1960s. Historically, this neighbourhood had been a neighbourhood of incredible struggles, in the Resistance [to fascism during the war], then later in the 1950s, and in '68, [there was] Sarca Avenue that exploded ... So here there is a fabric that is strong and very rooted, but that is absolutely disoriented by what has been happening in the last 20 years. It has lost all of the oldest reference points that it had. This space itself [Casa Loca], for example, was an old reference point, because it was the after-work building for the Pirelli factory workers. Here was the Pirelli cinema, an external courtyard, there was everything the worker needed for recreation at the end of the work day, so also symbolically at this level it was an old reference point. So we find ourselves on this thin balance line. (11 March 2005)

Giacomo stressed that it is especially significant that Casa Loca itself occupies a disused and abandoned building of the Pirelli factory. This company, in fact, was a major player during the years of Milan's industrial expansion. Following the closing of its factories, Pirelli rein-vented itself, among other things, as a real estate developer. As Manuel Aalbers describes, it emerged "like Phoenix from the ashes – a car tire manufacturer turned real estate developer using their derelict brown-fields for large and profitable urban redevelopment projects" (2007: 186). In this context, Pirelli's transformation is indicative of a more gen-eralized change in the use of the urban territory.

Here, like in chapter 6, I cannot but think of the abandoned after-work building as a spatial version of Avery Gordon's (1997) ghosts: the pres-ence of a significant absence, from where spaces can be redrawn and rethought, and entire cities recreated. By locating itself exactly in this zone of boundary, on the edge between not only historical periods and neighbourhoods, but also between uses and representations of spaces, Casa Loca might indeed be in a strategic position for reconceptualizing public space, knowledge, and identities. Giacomo's idea of creating a space on the model of the shopping mall is interesting in this regard:

You cannot put back into the field old tools [...] Some of the Social Centres continue to talk about the movement and struggles of factory workers but I think [...] they still live of projections that do not exist any more [...] The changes are more and more ... oh, how should I say it in a few words ... what is out in the field on the other side is very interesting ... obviously it attracts the attention of most people in a sweeping way. So you, in order to adapt to these rhythms, to the offer that your enemy has, you also have to use its languages, you have to use them to your advantage – perhaps by mimicking them [...] to show everyone how false they are [...] – but you have to be in the position to be able to use its instruments, you have to be able to study them [...] and give them meanings [...] So [this has meant] a distancing from what were the old styles of the left, according to me not [far] enough yet. It is obvious that you always carry history, and your past, within you – also rightly so – but [there is] a distancing that is more and more marked. Otherwise, your failure is certain. From the other side, you have an adversary that is too strong. For example if you consider that the piazzas, a symbolic place of sociality, now are empty, and instead the shopping malls [...] on Sundays people go there, even without buying anything, walking and chatting, that is a space of sociality. You have to study things like that. We in Casa Loca we have a project, for example [...] of rethinking this place in its characteristics like a shopping mall, using the same instruments but with our own offer. So, for example, to create a piazza that is covered [...] [using] a series of positive characteristics [that the malls have] [...] So we thought of covering our courtyard, making it into a piazza, and then around it a series of [alternative] businesses [...] a "Mall Loca," [...] a "Mall-titude." (11 March 2005)

Giacomo's idea of creating a "countermall" in Casa Loca can help us think about the forms that nostalgia can take (Berdahl, 1999). According to John Foot, the industrial past of the city is the source of romanticism and feeling of loss for many people of the left and for many workers (2001: 175–6; see also Muehlebach, 2012: 5). Foot argues that often the world of the factory itself is seen as a golden era, despite the fact that many commentators of the previous generations depicted it as a monster; however, according to Paul Ginsborg (2003), the workers who have participated actively in the struggles of those years remember with fondness particularly the feeling of empowerment, purpose, and hopeful possibilities.

We can perceive a tinge of nostalgia both in Alice's evocation of the battles of the Leoncavallo and in Giacomo's description of the working-class part of Sarca Avenue, with its strong legacy of resistance. At the same time, however, Giacomo argues that history, and the nostalgia it inevitably generates, should not blind us to the ways in which society changes and, with it, the dreams and identities of rebellious subjects. The dismissed areas, and the Social Centres with/in them, can no longer be the place where the traditional left clashes with the traditional right, or where workers talk about class consciousness. Rather, they are located in a thin line of difference that requires new subjectivities, icons, and models rather than the ones of the past.

Here, once again, the active engagement with and reinterpretation of spaces, with all the struggles and social relations they embody and engender, is a crucial axis of action. Indeed, this is the aspect that helps give such a heterogeneous net of people, places, and strategies as the *centri sociali* the form and feel of a movement (Dines, 1999). This comes even more to the fore if we consider for a moment that there are interesting discontinuities between the Leoncavallo and Casa Loca – indicative of the generational gap between the two and the different ways in which they have been positioned towards the social, cultural, political, and economic changes that have affected Milan and its territory. The very way in which I approached the two places is telling in this respect. When I was in Milan, I went to the Leoncavallo because, as almost everyone else in the city, I knew about it even before I visited it. I was well aware of its "communist" roots and reputation, of its role as a political landmark, and I was particularly interested in its exposition "The City That Will Come."

In contrast, I found Casa Loca in a much more circuitous way. Although I had first heard of Casa Loca from Alice, who offered to take me there one day, my actual visit was the direct result of having finally found San Precario on its doorsteps, after a search that took me several weeks. Alice herself was ultimately the one who revealed that San Precario, whom I had heard about from acquaintances and from reading about him in the communist daily newspaper, often lived on Sarca Avenue. Figure 21 depicts San Precario as he appeared to me on that morning of revelation, just in front of Casa Loca.

Characters like San Precario and the collaboration with the architecture students – aspects that made me particularly interested in Casa Loca – exemplify some of the new roles, engagements, and oppositional creativities that might be needed in order to promote a more equitable Milan. In this respect, young Centres like Casa Loca might be better positioned to

Figure 21. San Precario in front of Casa Loca. (March 2005).

carry out innovative countercultural interventions than the Leoncavallo, whose existence has been so strongly shaped by frameworks and idioms that are more directly linked to communist ideals (in all their variations).

At the same time, my approach to Casa Loca reveals some important continuities between the Leoncavallo and Casa Loca. The two Centres inhabit the same network, and they are connected through people and actions. The Mothers of Leoncavallo, like Alice, in particular are well respected in the wider community. It is thus not surprising that I could find Casa Loca and San Precario by following this thread. Moreover, and most importantly for my argument, the active creation of liberated, self-managed, public spaces is at the core of places like Leoncavallo, Casa Loca, and similar groups. In this respect, the young Casa Loca is literally the offspring of the older Centre. Although the relationship between Social Centres and the urban territory has been changing through time, space has always been and remains the central medium and result of struggle. The strategy of squatting is the most visible expression of this. Both Casa Loca and the Leoncavallo illegally occupy places that have been left over in the difficult deindustrialization process. Maintaining these locations means making visible many of the social and political strands that constitute the very territory of the city. By being at the epicentre of this active spatial knowledge, the Centre as public space has meaning – as a place where residents of the city become public actors (see Guano, 2002), people capable of engaging politically, socially, and culturally with the society around them.

The continuities and discontinuities between the two Centres point to some of the disjunctures in Italian neoliberalism itself. As Molé (2012) describes, Italy displays a mix of state-regulated jobs with a precarious and flexible labour regime, which causes deep cleavages between differently employed persons and between the working population and the state. And Ginsborg (2001) and Muehlebach (2012) point out the peculiarity of the Italian situation, which has seen, on the one hand, incomplete processes of neoliberalization and, on the other, a very fertile ground for it – due to the only partial establishment of a welfare state in the past and to the fact that, in Italy, the state has never been a strong presence in the first place. This all has resulted in "an Italian state composed of multiple social, political, and economic orders" (Molé 2012: 5) that includes, among others, strong client-patron relationships, enduring state regulations of some parts of the labour force, the "Catholicizing" of neoliberal efforts (Muehlebach, 2012: 15), and a dramatic shift towards privatized care.

The construction of a particular kind of public space carried out by the Social Centres is a way, finally, to visually disrupt the neoliberal city. The façade of Casa Loca in very vivid, bright blue paint with a giant Zapatista head on its side, in a context where the colour of buildings is highly regulated; this is a telling example (see Figure 22). Indeed, after it was painted, the website of the Centre displayed the painting of the building for a long time as a key action by the Centre. Similarly, the carrying of the statute of San Precario in procession during May Day rallies, its starkly theatrical appearance in a supermarket, its guarding of the door of Casa Loca in a curious juxtaposition between a guardian saint, a guerrilla sentinel, and a work of public art – these are not only symbolic gestures, but also active visual and performative interventions in the space of the city. Another oppositional Centre, the Stecca degli Artigiani, had in 2011 hung bicycles on lamp posts and at the mid-height of houses in order to lead people to its door. Significantly, when standing on the street and looking at the bicycles,

Figure 22. Casa Loca. (April 2009).

they appear juxtaposed to the giant cranes that are rebuilding the neigh-bourhood, a development that the Centre has been actively opposing now for several years (I discuss this more at length in chapter 6 and in the conclusion).

When listening to Alice, I was particularly fascinated by her dramatic and strongly pictorial rendition of marches and demonstrations and of the rebuilding of the Leoncavallo, brick by brick, by its supporters. In her narration, there was a sense that every reclaiming of space was a recreating of a presence in the city, by an organization that wants to be a visible mark of difference. If her long and well-rehearsed tale (I have included only brief excerpts here) was one that, it seemed to me, had been told many times already as an active claiming of history, it was also a strongly performative act that painted a picture of opposition. I had the same impression seeing the multitude of graffiti on the door of the Leoncavallo, and its banner stretched across the sunlit courtyard, and ready to be used in rallies. To quote it one last time, it read, "Return the city to us. We want it back."

Chapter Five

Creating Spaces, Constructing Selves

The actions, ideas, and experiments of Milanese Social Centres show that public space cannot be taken for granted. Significantly, the encounters I had with migrants living in Milan emphasized that public space is a precious resource that needs to be continuously claimed, made, and realized. When discussing this dynamic, moreover, it is important to keep in mind that the very relationship between public space, immigration, and multiculturalism has been a significant source of conversations in Milan. Indeed, one of the aspects that particularly caught my attention during my research was how often the "Italian" people I met talked about "foreigners," migrations, and multiculturalism when discussing – or walking in – public spaces.

In addressing this complex conjuncture, my goal in this chapter is twofold. First, I trace how my interlocutors who migrated to Italy from other countries described, constituted, and inhabited public spaces. Second, I show that public debates and everyday comments on public space and diversity not only simply refer to the many and different ways in which immigrant communities participate in the city. They also reveal some of the contradictory attitudes towards cultural identity, race, and ethnicity that are at work in Milan and more generally in Italy. A large body of literature has examined the difficult position of migrants in Italian and European societies; however, here I am interested in how this exclusion is articulated and negotiated spatially. In particular, I argue that public space, as a shifting yet significant category in town, acts as a pivot both for the voicing and/or enactment of exclusion and for the practising of forms of resistance against discrimination.

As Merrill (2011), Bonomi (2008), and Cole and Booth (2007) show, migrants constitute a necessary, cheap, and easily exploitable labour pool. Aldo Bonomi reports that in Milan their involvement grew from 5% to 25% of the labour force (2008: 9). In the construction industry, migrants comprised 40% of the workers in 2006, while they had been only 7% ten years earlier (ibid.). Immigrants' work is a crucial resource for Italian businesses especially in many occupations that are refused by Italians because they are too low-paying, dangerous, or socially undervalued.[1] This is the case, for example, with domestic work, a realm where non-Italian labour is central (Muehlebach, 2012; Cole and Booth, 2007). Immigrants make up a sizeable portion of the informal sector, which is an integral part of the Italian economy (Merrill, 2011).

Notwithstanding their important social roles, foreign-born (and especially visible minority) residents are perceived as standing apart from society and as irremediably different from "Italians" (Krause, 2001). Cast in the position of "non-persons" (Dal Lago, 1999), they are both materially and symbolically "locked outside of the Italian social-political community" (Merrill, 2011: 1543). According to Frisina (2006) and Pratt and Grillo (2002), this discrimination results, in part, from a profound ambivalence within Italian identity itself. Because Italy has historically been a divided country, struggling to create a national identity, outsiders of various kinds have been instrumental in drawing the boundaries of the community. The myth of the "nation as a coherent, monolithic whole" then posits non-Italians as "invading" subjects and immigration laws and regulations as necessary tools to defend oneself from what is depicted as an "immigration 'siege'" (Merrill, 2011: 1546 and 1551; see also Hansen, 2000).

This line of thinking is nourished by the approach to immigration at the European level. Since the Schengen Agreement of 1985, Italy – as one of the southern limits of the European Union and an easy country to reach – has been thrust into the role of guarding the European frontiers, and it has been repeatedly criticized for failing to do so successfully. In Milan non-European immigrants have largely taken over the position of southern Italians who were until recently seen as the subaltern "others" of northern Italy. The most salient line of difference is now increasingly between those who can claim an Italian identity described as European and Christian[2] versus those who come from "outside" this imagined geographical and cultural universe.

This marks a double-edged inclusion/exclusion. Alessandro De Giorgi notes that the goal for Italy, and Europe more generally, is not to simply "keep unwanted immigrants out, but to let some in and keep them under conditions of institutionally sanctioned subordination"

(2010: 160). This is part of a neoliberal reordering of Europe, which requires a flexible, precarious, and low-paid working class as a necessary counterpart to the privileged lives of those who have the social and economic capital needed for freely travelling, working, and settling within the European Community.

Since it is their vulnerability and lack of rights that make them into a key wheel in the Italian economy, ignoring immigrants as active residents and subjects is indeed a precondition for keeping them in a subaltern position. As the ultimate flexible, disposable labourers, they are as desirable as workers as they are "undesirable" as members of the polity and the city (Merrill, 2011: 1557).[3]

The 2009 debates on the role of culture in Milan, discussed in the previous chapter, provide an interesting example of this situation. Groups, institutions, and newspapers commented on the meaning and importance of culture in the city, yet both the people promoting more official perspectives and the countercultural front were largely ignoring the growing backlash against multiculturalism taking place in Milan. And if an unfortunate hierarchy was established between culture as high art, literacy, and erudition and culture as an active engagement with politics, history, and collective memory, the idea of cultural diversity never became a noticeable element of the discussion. This is because immigrant communities were not seen as making culture in either sense of the term. According to common understanding, the culture of immigrants is of a different kind – one constituted mostly by traditions, religious practices, and colourful ritual. Mirroring an historic, and profoundly discriminatory, division between the culture of Western and non-Western people, this kind of culture is often associated with backwardness and is seen as largely irrelevant to the active making of contemporary Milanese social life.

These attitudes are very important for understanding migrants' experiences of different urban spaces – as well as the reasons why public space as a category invited people's comments on multiculturalism, diversity, and immigration. Piazzas, streets, and parks are key locales where white, Italian-born Milanese encounter the "other." On a positive side, this can – and sometimes does – become an opportunity for exchange and encounter. Indeed, especially in the later periods of my fieldwork, I witnessed more and more instances of cross-cultural exchanges and mixed friendships. More often, however, immigrants' presence in public spaces is seen as a negative attribute of locations, either because it signals disorder, degradation, and danger[4] (Dines, 2002; Saitta, 2011; Melossi, 2003; De Giorgi, 2010) or because it is perceived as disrupting the local sense of place (Saint-Blancat and Schmidt di Friedberg, 2005).

In this context, visibility becomes a privileged medium for encountering and apprehending diversity. Public space is one of the sites where some of the people who think of themselves as "Milanese" perceive some residents to be "strangers" because they look "other"-wise. This can be a combination of skin colour, ways of dressing, and ways of using public spaces. What I find particularly interesting here is that the importance of vision and recognizing then amounts to an everyday negotiation and reflection on who belongs to the community, akin to Partridge's description of an embodied, "street-level bureaucracy" (2008: 667, referring to Lipsky, 1980), in which citizenship is ascribed onto others based on complex relations of identity, alterity and desire.

In this chapter, I discuss these dynamics by reflecting on the insights of some of my interlocutors. I will focus particularly on encounters with two women originally from El Salvador, who I will call the "Schuster Youth," and with Don Felice, a young priest who migrated to Milan from Mexico. These three persons point out that journeys of migration and the particular social conditions in which many immigrants find themselves – such as a difficulty in finding adequate housing, discrimination and racism, long hours of work, and social isolation – shape the ways they understand and enact public space.

At the same time, however, Don Felice cautions us against an easy identification of "ethnic uses of public space" that is often voiced in the media and in everyday local discourses. His words remind us that diversity itself is socially constructed, and that it is thus necessary to question nationality and culture as a fundamental character of our relationship with the city. We can be both insiders and outsiders, in different ways, in particular times and places. As a strategy and a guiding metaphor for this line of thinking, Don Felice described public space as a "sand dune," something constituted by people in different spaces and in different ways, depending on their social circumstances and journeys through the city. As he suggests, the shifting nature of this sand dune does not make it any less present or significant; rather, it calls for attention to the form it takes at a particular moment and to the winds of social interaction that shape it.

Places for Gathering, Sites of Judgment

I met the Schuster Youth in the spring of 2005, at the Schuster Centre, a Catholic parish and community centre[5] located in the outer rim of the city. The two women I talked to had migrated to Milan from El Salvador when they were in their teens, and one of them worked as a live-in domestic employee for several years. At the time of our conversation, they were both students in post-secondary institutions. One of the Youth was particularly interested in studying the history of the Milanese Salvadorian community, which she described as one of the oldest immigrant groups in the region. Both women have been very involved in the Schuster Centre's activities for several years. Indeed, our conversation (6 Feb. 2005) took place in a room of the centre, during the women's free time between the Mass we attended and the activities of the youth group they were involved with.

When I asked them about public space in Milan, the Youth immediately exclaimed that they had "more space *there*." The shrinking of space they experienced when they moved to Milan from El Salvador invests at the same time public and domestic locations and, because of this, is perceived as even more constraining. One of the women told me:

> I remember the first time that I arrived in Italy, we were in the house of a friend and I asked my mother: "Mum, where is the other room? Where do we sleep?" Because I could not conceive the idea of […] a space so little. This is something also for the kids, for example, who come from El Salvador: […] they are used to big spaces, to run, even if in poverty, but there is space; you see, [in] the country one can move, one can run, one can scream, you are in your house, you can listen to music. Here you cannot do that, because you are disturbing the one that lives under, the one who lives over.

For the Schuster Youth, a sense of being watched from all sides and never quite having enough space seems to be pervasive in Milan. One of the women, for example, recalled with nostalgia sitting on Salvadorian town squares with friends, and compared it with having to endure hostile gazes and gestures in crowded buses. Similarly, other migrants I met during my research expressed the feeling that Milan was like a "prison." A study conducted by Naga argues that often "work acts on the migrant body like a mechanism of imprisonment" (2009: 8) because rather than bringing dignity and autonomy to the employed person, it furthers a dependent, subordinate, and vulnerable position. This is particularly the case with domestic labour (see, e.g., the comments below of one of the Schuster Youth on her work as a live-in caregiver).

The feeling experienced by many migrants of being controlled (see also Van Aken, 2008) is amplified for those who sell on the streets or who might not have official papers, due to the police checks in everyday life. My fieldnotes for a day in April 2009 can give a glimpse of some of these occurrences:

From the Duomo Piazza I walk through Via Dante. It is a beautiful day and many people are promenading in the central avenues. Street vendors, most of them young Black men, are peddling their wares. After walking about half of the avenue, I notice that there are groups of people that have assembled, all of a sudden, in the middle of the sidewalk. They are holding folded cardboard pieces, bags, or gathered cloths. They are glancing nervously around them, their feet ready for running. I realize that there is police around and that those are the vendors who, just a minute ago, were offering purses, t-shirts, and belts to passers-by. Some are huddled on the stairs leading to the subway entrance, and seem unsure whether they should return to the street or quickly get out of the way. I am struck by the stark contrast between those who are on the run and the rest of the people on the avenue, who can go on strolling in the sun notwithstanding the policemen walking up the street, and the worried looks of the groups of men.

A little later, as I reach the Sforza Castle, an even more dramatic scene ensues. Directly in front of the large stone entrance, there are other stalls on the pavement. All of a sudden, a car drives up the sidewalk at alarming speed. While the passers-by move out of the way to avoid getting hit, all the vendors abandon their merchandise and escape. Their movement resembles a wave lifted by a strong and sudden wind. The police agents who descend from the car pick up all the items that remain on the pavement and throw them roughly inside the vehicle. Leaving the scene, I walk around the Castle walls and eventually enter the park at the back of the Castle. Almost an hour has passed, yet amid the grass, the trees, and the children playing, the raid against the vendors is still under way. Two police cars are parked on a small path of the park, and about twenty young men are walking and/or running in the opposite direction, talking animatedly to each other. (13 April 2009)

Although scenes like the ones I describe here do not happen every day, they occur often enough to bring an element of fear into the lives of many migrants and to strengthen the pervasive sense that the latter are "illegitimate" users of the city (Merrill, 2011: 1557). That these actions by police are also dramatic performances (Herzfeld, 2009), moreover, is an important component. They have the effect of separating city dwellers, people using the same piazzas and streets, in two very different categories. Black bodies, the most frequent targets of raids, are even more strongly designated as migrants and potential criminals, and white bodies are inscribed as the innocent audience of this performance, who can continue strolling as an embodied sign of national belonging (Fikes, 2009).

Even if they seem less "inhabitable" than the ones in El Salvador, the Schuster Youth explained that Milanese public spaces are nonetheless significant locations for gatherings, recreation, and for their participation in the daily life of the city. Indeed, for them, one of the most important aspects of public spaces is their ambiguous role as both a resource and a burden – as both places of freedom/sociality and places of judgment and control.

In our conversation, the two women talked particularly about Milanese parks, the Duomo Piazza, and the Schuster Centre. The latter, they explained, is especially important because it works as *the* main gathering site for the Salvadorian community – who thus make it into "their" central public space – irrespective of the fact that it is owned and regulated by the Church. Here it is important to note that the women described the Schuster Centre as one of the most significant public spaces in their lives because it is a community space for a specific group of people, rather than being an open and publicly held location. One of them said:

Sure, [here] we [still] do not have that [much] freedom, because […] it [the Schuster Centre] is not ours, […] but, for now […] we succeed in doing many important, big, celebrations here […] Like for example the […] independence from the Spaniards is celebrated here. So for us it is important to have a space. Not all immigrants have a community with a space, so this is the result of people who have worked earlier, who have fought for having [a space].

The Youth discussed public space in relation to their membership in a community with which they strongly associate. Their comments echo the observations by Mauri and colleagues that Milan des not offer many spaces where immigrant groups might celebrate festivals, hold religious ceremonies, and more generally, meet and socialize (2003: 231). Indeed, the women explained that Salvadorian-Italians are among the

few lucky ones to have a place at all where they can gather and speak
Spanish. Even then, and similarly to other communities' use of church
spaces, the Schuster Centre is a locale that they can never entirely call
their own. This location, moreover, might distance from this community other Salvadorian-Italians who do not wish to be connected with
the church.

The general lack of meeting places for immigrant communities is
due to the expense of renting, buying, building, and/or maintaining
a cultural or community centre, as well as to the scarcity of existing
community structures that are not affiliated with local religious institutions. Some immigrant groups with a Catholic background, such as
Salvadorians in this case, can use existing pastoral buildings. However,
communities that have little or no affiliation with Christianity or that
wish to keep religion more in the background may have to use sites like
post-industrial facilities for gatherings and festivals (Granata, Novak,
and Polizzi, 2003: 100; Novak and Andriola, 2008). Especially for Muslim minorities, moreover, the resistance of neighbourhoods to the visible presence of "foreign" residents can become a very difficult obstacle
to establishing a community space (Saint-Blancat and Schmidt di Friedberg, 2005).

With regard to open public spaces, the dynamics are even more complex. Because Milan has very few parks, because its streets and piazzas
are usually busy and crowded, and because it has been until recently
a quite homogeneous society, the use of public spaces by immigrants,
whether as individuals or in groups, tends to be seen as posing a challenge for the "co-inhabitance of urban spaces" (Mauri et al., 2003: 231).
Picnics in the parks and gatherings in the Duomo Piazza mentioned
by the Schuster Youth are interesting examples. According to the two
women, parks are important sites of gathering for Salvadorians in
Milan, because parks allow for many social activities to take place at
the same time. One of the women explained:

> Public space for us [...] is the one of gatherings and celebrations, music,
> being together, for this reason it is often a park. Especially during the good
> season, we organize ourselves, sometimes even with barbeques, and all
> that, and [a park becomes] a meeting point, where perhaps one plays soc-
> cer, [...] softball, basketball. So it is a way to meet, to eat together.

Don Felice, too, described visiting one of the parks and finding a
group of people who reminded him of his country, Mexico, listening

to music, selling food and small items, and creating, in his words, a wonderfully festive space:

> There is another space [...] that for me is very interesting [...] It is a park [...]
> Especially Peruvians meet here [...] you can go around the park and here,
> for example, [...] you can find lunch, here you can find music ... and here
> only games [...] It is a family place, the problem is that there are so many
> who come here every [...] Saturday and Sunday to meet. According to
> me, [...] because they cannot find space perhaps in this area [the centre of
> the city] let's say European, with European character, they gather, [they]
> construct a space for themselves that relates to the memory of their cul-
> ture [...] Whenever I have been here, I have been immediately transferred,
> transported to my culture, in the sense that the music, the salsa, the cum-
> bia, [...] the Latin-American aspect you can breathe it immediately [...]
> This is another space [...] that people construct for themselves. (5 Nov.
> 2004)

Picnics in the park held by visible minority residents, however, are not always regarded with sympathy. The Milanese section of the *Corriere della Sera* of 7 May 2005, for example, writes that "Milanese parks are off-limits for the residents," because on weekends they are "'occupied' by foreign communities" (Verga, 2005). In a similar article of 2009, a representative of the Northern League decries that the use of green spaces by immigrants has been creating significant tensions in Milan every year, especially with the arrival of spring and summer (Indini, 2009). And Mariangela Giusti confirms that many white, Italian-born residents increasingly see in a negative light immigrants' recreational activities in parks, thus perceiving parks as "closed" and no longer accessible spaces (2008: 19).

The issue, according to these articles, is that gatherings of immigrant groups in the parks are too loud, last for too long, ruin the grass, leave the space dirty, and often even break the law, as when the people drink alcohol there or take money for tickets for dancing or listening to the music. Conflicts around park use have occasionally even erupted into violent clashes, such as the ones in July 2007 at the Cassinis Park (see Gagliardi, 2007). These tensions involve most parks of the city. As Rossella Verga poignantly described "we hear disquieting signs even in the very central Public Gardens of Porta Venezia" (2005). The spokesperson of an association created to "defend" one of the parks explains, "We only want respect of the rule and of nature [...] not discrimination. We don't hold anything against the foreign communities, but if we wanted to say it provocatively, we could say that there has been an ethnic cleansing in the other direction. These days, the inhabitants of the area on the weekend go somewhere else" (ibid.). Another spokesperson quoted in the Verga article similarly declares, "The citizens, on the weekend, do not go any longer to the park." Echoing these statements in 2008, a resident complained that immigrants gathering in parks have "transformed the [...] street into 'a nightmare for the inhabitants' [who] on the weekend hope for the rain to come, so that the immigrants would stay in their homes" ("Ripamonti, via Coari," 2008).

Particularly interesting in the articles that appear on this topic is the way in which they contrast "Italian" and "foreigners" regarding using parks, and the way in which immigrants become, by default, not "residents," not "Milanese," and not "citizens." Significantly, the use of parks emerges as an issue of visibility, presence in, and entitlement to Milanese spaces. Deputy Mayor Riccardo De Corato is cited in one *Corriere della Sera* article as saying, "in Milan, apart from the illegals, there are over 150,000 regular extracomunitari who work all week long and who cannot disappear during the weekend. And while many Milanese have the good fortune that they can go to Rapallo [on the seaside] or to Courmayeur [in the mountains] on Sundays, they do not have any other meeting place than the parks" (in Foschini, 2005). Here the visibility of immigrants in parks does not lead to their greater participation in society – according to the idea that public space makes visible subjects and groups in order to facilitate their representation in the public sphere. Rather, it seems to spark a (hidden) desire that they could just "disappear" on Saturdays and Sundays, leaving the city for its "residents," the invisible Milanese who go somewhere else than the parks.

De Corato's words rightly show that particular people use the city in particular ways, depending on a variety of economic, cultural, and social factors, and on social positions such as class, race, nationality status, and gender. However, his comments still suggest the imaginary of two different, monolithic, and internally homogeneous publics – the Milanese and the *extracomunitari* – who are entirely different from each other.

Here I would like to point out that Don Felice and the Schuster Youth, too, talked about "immigrant communities" and "Italians" and their uses of public spaces. After all, there are culturally and locally informed ways to do things in places. These, however, always intersect with particular contexts, practices, and structures and cannot be understood as an essential and foundational relation to places. Importantly, the Youth and Don Felice emphasized the particular, social conditions of the people who migrate, their specific journeys of migration, and the processes by which these Milanese are *rendered* foreigners and rendered *extracomunitari* in Milan.

The Schuster Youth talked about their identity as Salvadorian migrants to stress that their social position in Milan affects the "claims they can reasonably put forth" (Tsing, 2000: 338) about their role, entitlement, and participation in urban public space. On the contrary, in much media and political discourse, immigrants can only and necessarily relate to public space on the basis of nationality, race, culture, and ethnicity. It makes it possible to say, "Oh, look, these are immigrants, they are using the city in an immigrant way."

"Non-Italians" living and working in Milan resist these lines of reasoning in several ways, through the work of associations, initiatives, and everyday, individual commentaries. The following anecdote from my fieldnotes is interesting in this respect.

In the large empty space of a subway station, in the passage between its two entrances, there is a group of seven youths. It is five o'clock in the afternoon. They speak Spanish to each other, they have brought music, and are starting to dance. When I ask them what they are doing and why, the youths tell me that they meet every day here because there is space to dance. One of them quickly and urgently adds: "Also the Italians come here." (25 Nov. 2004)

The youth's last sentence, and the way it was spoken, was a counterargument to local discourses. What he was indicating is that this group of friends dance every day in the subway not because they are not Italians, and thus necessarily use public spaces in a different way from how Italians do – but because the unusually large smooth floor of this part of the station makes for a wonderful dance floor, which is at the same time free, accessible, and warm in the winter.

In a similar vein, during a conversation at a bus stop in February 2009, four North African–Italian women emphasized to me that they want to be considered Milanese just like everyone else living in the city. They explained:

"You know, we work, [thus] we are Milanese."
"Yes, we are Milanese, for many years," [her companions repeated in agreement]. (3 Feb. 2009)

In our conversation, the women especially seemed to refute the connection between their visible minority status (as Black women wearing headscarves) and the label of "outsiders." For them, it was the role of workers that made them an active part of Milanese society. The comments of these women reveal the widespread slippage between visibility, ethnicity, and nationality. Because white, Italian-born Milanese often attribute the status of "immigrant" on the basis of the "other's" appearance as a visible minority subject, and by the way she or he is seen to inhabit piazzas, parks, and streets, the fact that many of those called "immigrants" are Italian citizens or were born and have lived all their lives in Italy[6] is effectively excluded from the conversation. Importantly, this has the effect of naturalizing citizenship as something that is commonsensically attached to a white person, who represents the racialized national body (Fikes, 2009).

In the Duomo Piazza

Similarly to parks, the Duomo Piazza, according to the two Schuster Youth, is an important public space for many migrants, while also being in many respects a place of contention.

> YOUTH: Unfortunately if you work [as a live-in domestic employee] in a family and you have perhaps the Saturday and the Sunday free, you do not have a reference point. So the park becomes a meeting point [with other Salvadorian-Italian people]. However, in the solitude you find yourself in, far away from home, perhaps without a job, perhaps you start to drink. So you see, for example, in the Duomo Piazza in Milan, it is full of immigrants because they do not know where to go, perhaps they do not know that there is a community, in our case a Salvadorian community. For this reason [the Duomo Piazza] becomes a gathering point [...] And the Italians say: look at all the foreigners with the beer in their hands [...]
>
> CRISTINA: So in the Duomo Piazza go mostly people who [...] find themselves lonely, less in contact with a community?
>
> YOUTH: Yes [...] Have you gone to the Duomo Piazza? Try passing there on a Thursday. Because it is the free day [for many live-in domestic workers]. Or on Saturday. Or on Sunday.

According to the Schuster Youth, the Duomo Piazza is a *necessary* place for many Salvadorian-Italians who "have nowhere else to go."

This speaks to the fact that public spaces are often the only thing that is left to people who do not have access to adequate private or community places and/or services (Bayat, 2012; Mitchell, 1995). In this case, because of the social isolation and harsh labour conditions of many immigrant workers, the Duomo Piazza is an important site of sociality and recreation. As mentioned by the women, this raises complex debates on proper versus improper uses of urban spaces and their effects for the residents at large. Still, and in spite of the women indicating that only more socially isolated persons use it, as a vital meeting space, the Duomo Piazza becomes a site of social interaction, allowing the people who use it to claim a space in the city. Don Felice's comments on the Duomo Piazza emphasized this positive aspect:

> This one [the Duomo Piazza] I would rename it Piazza of Cultures [...] Especially Latin Americans are slowly – how can we say? They are transforming this space into their own space, [...] a place of their own which, however, always remains open [...] Near this one, there is [...] San Fedele [...] It is interesting, because San Fedele is the [...] piazza of Philippines. Because here you see many Filipinos, and Latin Americans [...] Especially the teenagers are slowly slowly starting to construct this space for themselves, as a piazza, as their own space [...] We can say that they are appropriating this space [...] And the Nigerians [...] they start to sell books [...] They too are starting to be very much in this space [...]
>
> This [Duomo] piazza is filled with many ... can we say cultures? [...] Every day I see a Latin American who sits here; on the weekend they meet to make the space [...] It is as if it was an open space.

Similarly, Don Felice recounted that near where he lives, a group of Albanian-Italians meets every day at the same time in a little park to play soccer. According to Don Felice, that park,

> that is already itself public [but that] no one from here, the Italians, would use, would play on [is] inhabited by these people [the Albanian-Italian soccer players].

In this way, the park is *becoming* a public space.

Streets, parks, and piazzas, however, can easily become loci of scrutiny and judgment. Although this holds true for all city dwellers, the

Schuster Youth indicated that it is especially the case for people who are usually identified as "non-Italians." Their description of the Duomo Piazza reminds us that immigrants tend to be seen – rightly or not – as group users of public space more easily and more often than white, Italian-born inhabitants are. According to Mauri and colleagues, in fact, one of the questions that emerge in the context of immigration and public space is "How is a coexistence of different uses, *of collective versus individual and family uses* of open spaces possible?" (2003: 231, emphasis in original). This question sparks many other interesting ones: how do we define what collective versus individual uses of public space are? Who engages in which one, in what contexts, and why? And when is a person who is moving and participating in the city an individual and when part of a collective? Similarly, Valerie Amireaux (2006) asks when is a Muslim immigrant in a piazza a Muslim and when is he or she a citizen? Which signs and practices can distinguish one from the other, and who decides on these criteria? Is a veiled woman in public space always part of a collective identity? Is, on the contrary, an unveiled Muslim just an individual woman?

These questions alert us to the unseen, taken-for-granted, and normative whiteness of public space. A group of "Italians" in a public space are generally seen as "people in the street," but a group of, say, "Latin-Americans" is often perceived as a "noticeable group of immigrants." Mainstream discourses that attribute unexpected uses of public space to different ethnicities and nationalities both result from and nourish these associations, as they help construct public space as inherently European and white, and the uses "others" make of it as changing and reinterpreting it.

These ways of thinking emerge in both positive and negative versions. A negative one is the common complaint that immigrants are invading public space (see, e.g., Dines, 2002; Maritano, 2004). A positive one, which however equally essentializes identity, is the comment I heard from several people that "Italians" do not use public spaces anymore: they simply consume and/or window-shop, while "others" "really" use them. For one thing, this discounts window-shopping as a way of using and claiming public space; for another, it romanticizes "non-Italians" as people who "still" veritably enjoy public space.

In opposition to this, the Schuster Youth indicated that it is more fruitful to attend to the specific roles that spaces play in relation to the different social positions of their users. The description of one of the women of her free time in the city, in which she can get a break from the constraining work as a live-in domestic employee, is particularly telling.

YOUTH ONE: Perhaps there [in Duomo Piazza] you can find a co-national, even – you have to understand – even the simple fact of being able to speak Spanish, rather than ...

YOUTH TWO: ... working [...] because I have had this experience [...] I worked for four years with this old lady who was nice and all, but, indeed, you feel like you are in jail. You have to ask permission to go to the washroom, to take a shower, ... if you eat, she has to be there to see what you eat, and if you do not eat she is still there: why didn't you eat? You are not well? But one is well, is just that [...] sometimes one feels [...] sometimes I remember I would say: how beautiful it would be, if perhaps, even if not my mother, but at least someone else was here, with whom I could talk about something. I could speak Spanish, and so I remember in fact every Saturday when I went out I felt free, really free, to exit from the front door – how wonderful! I went out all night long, with the friend I lived with [...] I talked all night long [...] all of those words [that were] inside me that perhaps [otherwise] I could not say to anybody [...] And there I also remembered that [...] I come from the countryside, right [...] I remember that at home in Salvador ... the house is big [...] nosotras traiamos las sillas y las poniamos en la calle, that is, we put the chairs on the street [...] and there we would sit and chat, how beautiful, and there on the other side, close by, there was a piazza, there were all the kids playing [...] but everything was beautiful, everything outdoors, and here instead you cannot afford [are not able] to do that.

YOUTH ONE: ... also the time [here is lacking] [...] there is also a change of climate [...]

YOUTH TWO: ... but sometimes [...] sometimes there is perhaps a little hour [...] but [even so] ... here, there is no space to do that [...]

YOUTH ONE: You don't know where to go, you do not know what to do, because, we can say, society does not allow it, or the place does not allow it, it is many things [at once].

More than a fixed and definite place, public space here emerges as a dimension across time in which to experience freedom and human connections. This causes this interlocutor to switch languages and to resituate herself in a different place, where past, present, and future might exist without interruptions. Most significantly, the women's insight is that the very meaning of space is always deeply dependent on one's experiences and life stories in the city. One of the two women summarized this in a very poignant way:

> She came here alone when she was 17 years old […] I came here alone with my mother when I was 11. We have been each other's [only] family for all this time. So now [that you know this], do you understand why it is so important to us to have a place where we can meet [other Salvadorian-Italian people]?

The woman's comment was phrased as an urgent reminder because it was directed to me as a person who could clearly rely on family connections in the city, and who could easily claim Milanese public spaces as my own. My interlocutor rightly assumed that I would need to be able to imagine what it might be like to live in Milan alone, or with just another person, to work long hours, to miss my first language, and to be in a precarious situation, to even just start to understand what particular public spaces could mean for people who are differently positioned than myself.

The Schuster Youth indicated that it is not possible to understand the meanings and forms of public spaces, nor the way in which they work, without considering the social positions, the personal journeys, and the circumstances of those who use them. For many migrants who work long hours and cannot have a space of their own, parks, piazzas, and streets are a crucial resource. Not only because they are a space outside of one's house – which for many immigrants is often too small, or is the realm of their employer – but also because they offer the opportunity to meet with others.

Unfortunately, even if public space is often a necessity for many immigrants, the latter are often those who have the hardest time to claim a space in the city. Indeed, this seems to be almost an opposite situation to that of the VivereMilano's participants, who can easily perform, inhabit, and constitute an agora in the piazzas and streets that they have ignored and escaped from for many years.

Here I would like to turn to Don Felice's words. Echoing the insights and concerns of the Schuster Youth, he reminded us that the ways in which people consider themselves and others to be Milanese/Italians and "strangers" complicates ways of thinking about public space. Particularly compelling in his commentary is his subtle substitution of imagination for immigration, and creativity for cultural identity, which invites us to think critically about the construction of both self and others in urban locales.

Public Space Is a Sand Dune

Don Felice moved to Milan from Mexico to study theology at a Milanese university. In addition to being a student, Don Felice had been officiating as a Catholic priest in a parish close to the southern periphery of the city. Shortly after I left Italy in 2005, Don Felice moved to a church in a smaller community outside of the Milanese municipality, joining the many commuters into the city in order to continue his studies in central Milan. I met Don Felice several times during my research, in 2004, 2005, 2009, and in 2011, and in a variety of settings. These included the streets and cafés of the centre, my apartment, and his parish, where I attended the Mass he celebrated and where I took part in a Mexican gathering and celebration he helped organize for the congregation. In what follows, I present some of his insights on public space drawing from two of our encounters: an hour-and-a-half interview that took place on 5 November 2004, in the church office where he lived and worked, and a two-and-a-half hour guided walk in the centre of town, on 18 November 2004, during which we met several other people, and visited the university he was attending at the time.

In my writing below, I decided to combine these two very different conversations, contexts, and texts, because what I learned from Don Felice resulted not only from what he told me during our sit-down interview, but also from the walking conversation we undertook together. In the latter, public space became visible as a shared context that allowed words to be spoken and listened to, images to be seen and interpreted, encounters to take place – and all of those things to be woven together in a meaningful fabric. This walk through the city was for me a reminder of the ideas and concepts that Don Felice shared with me during our previous conversation about public space.

More precisely, an important part of what I learned from Don Felice emerged from the very difference between these two encounters and the two texts that I wrote to represent them. The fact that the sit-down interview was taped and the walking tour was not brings this particularly to the fore. Although I had brought my tape recorder with me during our walk, there was no moment in our city journey that seemed stable or settled enough to start recording. This is because our conversation was interrupted and enriched by many encounters, which weaved in and out of both our movements through the city and Don Felice's words. Images and sounds became important parts of the conversation.

Rather than putting this encounter aside as "too messy" or too fragmentary to say anything about public space, I include it here because its very "messiness" is exactly what can give us a sense of the character of public space according to Don Felice. I thus created a composite text of the two conversations, in which the sit-down interview text is written in regular characters and the walking tour in italics. My aim in doing so is not only to include both conversations and encounters, but also to highlight what lies between them: their links to and differences from each other. As Fabian (1990) argues, in fact, sometimes ethnographic knowledge comes from a "missing text" or from the gap between dialogues. The very difficulty in "pinning down" and representing fluid social interactions in and through spaces (which came to the fore during our walking tour) is a perfect illustration of Don Felice's interview description of public space as akin to a river that is always flowing and always new. It is indicative of his insistence in using metaphor to represent something that, because it is so ephemeral and interactional, is especially hard to explain and define.

Before turning to the text, I would like to point out that some aspects of Catholic theology inform not only Don Felice's ideas on public space, but also and more subtly the very practices of its construction that we both employed during our walk. In the Catholic thinking that has a pervasive influence in Italy, public space is significant as a site where people can realize their community with each other guided by the ideals of tolerance, brotherly/sisterly love, charity, and redistribution.[7] It is important here to note that this does not necessarily correspond to the actual practices and roles in Milanese society of most of the churches, parishes, or religious communities – let alone the institution of the Catholic Church, which has rather promoted hierarchy in city spaces. However, Christian philosophy was a source of idealized conceptions for many of the people I encountered, including several Terre di Mezzo members and the Schuster Youth.

More specifically, during our walk, Don Felice told me that public space is based on "the word." To understand this, we have to remember that "the word," as communication, enunciation, storytelling, and dialogue, is seen in Catholicism as a transformative, productive, and creative force in society. This aspect was quite significant in our walk. If our itinerary did conjure public space and enacted it as a fleeting

presence, it was mainly through communication as a creative and productive moment of encounter. That is why dialogues with people along
the way were an important part of our walk. They were part of the
creation of a temporary community, which could be the basis for realizing public space. This way of building public space contributed to the
idea of the latter as unstable and shifting, processual and performative.
Although I found this way of approaching public space illuminating,
this focus on the "word" might make it harder to see other ways of conceptualizing and/or enacting public space – such as the one of Social
Centres, which can be better described as a flexible heterogeneous network than a clear and shared community, and whose view of space is
centred less on the role of communication between subjects who can
speak the same language and more on political conflicts and sustained
historical struggles.

> DON FELICE: One time [...] another Mexican friend of mine came here, who
> had been here eight years ago. And I tell you: we went walking through
> the same streets that I always take [...] and he told me: "Look, I am sur-
> prised, because I got to know a Milan that I did not know." In the sense that
> he started to know, for example, the space of the Naviglio [...] or also the Brera
> area, or even San Sempliciano that is also very beautiful, or this other side
> [...] But I tell you: I still have to learn so much about the city, but [I learn
> about it] through the work that I do, or the journey that I do every day,
> and the way in which I relate to the city, both by going through its open
> spaces, and by going underneath it ... in the subway ... because for me it
> is also interesting to get to know the city under Milan, [...] because [...]
> it is a great majority [of people] [...] who move underneath. And even there,
> I find very interesting [how] public space constructs itself: with the news-
> paper stand, the café [...], the small booth to take photos. Also all the peo-
> ple who come to sell – someone [selling] an umbrella, someone a CD [...]
> The space underneath is, according to me, a mobile space. [For example]
> the café usually does not have tables, no [...] it adapts itself to this space
> [which] [...] is always a space with people who pass by [...]

*I met Don Felice at the metro station, and from there we walked to the Fac-
ulty of Theology where he studies.[8] When we arrive to the Faculty, we enter
the bookstore [...] There Don Felice sees a friend, a priest from Ghana. He
introduces me to him, and explains that I am doing research on public spaces.
"Perhaps one day you could talk?" He suggests. "He has been in Italy longer
than me, he can tell you about public space" [...]
From the bookstore, we pass through a door that leads to the other part of the*

building, where lessons are held. I hesitate: I am not sure that I am allowed to go in, since I am not a student of that school. This is not supposedly a public space! But Don Felice urges me in, and there we are: in a beautiful monastery, with its cloisters, the hallways which smell old and grand and are adorned by paintings and sculptures, the grand stairs that are so incredibly quiet. Women and men are walking, or are sitting at the desks reading. We walk into the courtyards, which, as Don Felice tells me, used to be the monks' gathering spaces. I take pictures, although they do not seem very meaningful when I look at them later at home.

We meet another colleague and student of Don Felice, Don Pietro, who is here from Togo. Together we walk though the hallways, through the huge doors made of dark wood and glass, and back out into the street.

> DON FELICE: Look, every place I have known it through a person. I think that this, too, is a way of appropriating the city. [Another way] [...] is walking by myself to discover [...] I, for example, always try to invent a new route. But the first approach [to a place] is always through a person.

We walk to a café nearby, where Don Felice, Don Pietro, and I sit at a tiny table with sandwiches, coffee, and apple cake. Don Pietro tells us about his work in a harsh neighbourhood in the periphery of Milan. He explains that many immigrants from sub-Saharan Africa he knows do not have time to use public spaces, because they are so busy working in small factories, juggling several employments.

After a short time, a man around 50 sitting at the table beside us joins in our conversation. Seeing the apple cake I brought with me and that I am offering to the two priests, he asks: "Is that apple cake? I love apple cake." "I am ___," he laughs, introducing himself. "I am an art restaurateur."

Don Felice spins this at once into the issue of public space we have been debating: "If you are a restaurateur, tell us: how can we restore public spaces when they do not work?"

Everyone laughs, and nobody answers his question.

We leave the café, and promise to come back with more apple cake.

> DON FELICE: One time, I remember, I was looking for a CD. I had seen it the first time that I had been here [in the Naviglio canal neighbourhood], at around 11 at night [...] That other day I went to look for [the CD] at around 9 p.m. and no one was there [...] I came back later and [the CD sellers] were all

back and I saw that after 9 p.m. this neighbourhood slowly starts to reconstruct itself, [it becomes] a small road in which not only CDs but many other things are sold [...] According to me the people who go around here, walking, according to me they are not ... not many of them live around there. They all come from outside [this neighbourhood], they all come just to walk here [...]

[P]ublic space [...] constructs itself as a space which now it's here and now it is not. For this reason, I was telling you that these people [the CD sellers, the people who come to the small markets to see and buy], you find them in a place, then in another. It seems to me that in this way the city itself [...] starts to move, it is mobile in this sense.

[...] And because of this, [...] [the neighbourhood, and the public space that is created] becomes ... how can we say it? We could say it becomes part of the river that passes through this neighbourhood. I think about [public space] like this: it's like the river, like the water. The movement is always new, because [...] whenever I go there I have never found myself with the same people, it is always new people [...] it is a public space which is not formal, is not established, that changes [...]

[In this neighbourhood] people meet outside, walking. Indeed, you see many people talking, but they never stop walking. One goes walking, goes to get an ice cream ... one starts walking, circulating, promenading, it is difficult to find [somebody still], there are cafés and bars but people generally [move] [...] this is a sign that it is the people themselves who create public space.

Out in the street, Don Felice starts talking about public space again. "Everyone seeks something in public space," he tells us. "I return to a space because I find what I look for." On our way, we pass by a busy market.

I am fascinated by how the market effortlessly enters the conversation of the two priests, becoming a living, embodied example of what Don Felice was saying at other times, in other conversations. I am spellbound as I cross that edge from where public space is no longer a description and an argument, but rather a movement, a walking that I can remember, and/or a moment of poesis, and performance that also interpellates me affectively.

Pointing at the market around him, Don Felice explains:

In Mexico, in regards to markets, and other public spaces, we say this. [Usually] the earth, the asphalt is cold. But where there is a market, a public space, a space where people get together, the asphalt becomes warm.

As we make our way through an incessant river of people through the piazza, through the stalls, he continues his talk:

> *Public space is a sand dune,*
> *because it is always made and always changes.*
> *Not only for youth, [but] for different subcultures.*
> *People move and meet*
> *in places and ways that are always different,*
> *and [so they] construct different spaces.*
> *Public space gets created in different places,*
> *in ways that are always different,*
> *it is fluid*
> *like people are who are moving.*

As we go on talking about public spaces, and the importance of dialogue, the two priests stop and look at a huge advertisement posted on the wall of one of the buildings surrounding the piazza. It shows an elegantly dressed man holding a stick and drawing a circle in the sand around where he is standing. The caption says: "I wanted a bank constructed around you."

The image enters in the conversation just like the market before it: Don Felice and Don Pietro say that many people want the world to be constructed around them, and serving only their narrow immediate interests. And then, they seem to imply, how would society, and public space itself be possible?

Suddenly, I realize that I have been in this exact same piazza a few weeks ago, with Francesca [see chapter 7]. What she had pointed out to me of this piazza on that occasion was not the market (it had not been there on that day), nor the advertisement, but a historic building. I am shocked by the realization that a different context and set of ideas had made the piazza almost into a different place, which I could not recognize as the same one where I had been with Francesca.

DON FELICE: I can tell you about my living in the city [...] I have a veritable desire to appropriate these spaces, also to live better, no? In the sense that, at the beginning, it seemed a foreign and very strange city for me. Little by little, however, every day I learn something new and this also helps me to appropriate its spaces and already it becomes for me a city known, and dear, so that I do not feel a stranger [to it]. But still, the strangeness remains within me, because you are always in fact a stranger [...] And this is also a topic that I am very interested in: how does one live as a stranger inside the city? [...] And this is interesting because this feeling determines also the way in which one lives and moves through the city, because, for example, I as a stranger, it is very difficult for me to enter a restaurant because immediately people notice their own gazes, the marginalization, the exclusion that they [themselves] do [to me] [...] Even the way they behave towards you, because you feel it at once.

And I tell you: even the people of the city experience being a stranger [lit., live themselves as a stranger] in many of these spaces. You can find yourself with people who tell you: "Oh look, it has been so many years since I have come walking in this part [of the city]. I always ... my space is this: [...] the house, the church." [They live] a more sedentary life, more quiet, more in their house, and they almost do not go around [in the city]. And if they go around, it is always the mountains or the seaside.

Strangers to the City

Don Felice's words and our walk above emphasize that public space is continuously made by people's movements, interactions, and performative enactments in and of urban spaces. His story of the daily awakening of the streets by the river, with their continuous promenading, their market stalls, and the movement of people, each of them looking for something in the spaces of the city, elucidates beautifully the daily manufacturing of public space. Indeed, our very walk in, through, and across different spaces – a walk porous to other's people's comments, images, events, memories, and recollections – is a living example of this way of conceptualizing public space.

His focus on the incessant construction of public space in daily life resonates with scholarly discussions on "representational," or lived spaces (Lefebvre, 1991). As a strategy to unravel and critically examine space, several authors have introduced a distinction between space as a material, produced location and space as a lived terrain, embodying diverse uses, histories, and resistances. De Certeau, for example, distinguishes between *place* as material, hegemonically structured locations, and *space* as the daily, lived embodiment of spaces by people enmeshed in the workings of power. Almost echoing Don Felice's narration and tour, he represents *space* with what he calls the "long poem of walking" (1984: 101) and illustrates how people living and embodying an urban place appropriate spaces, perform in and through them, and create relations between people and locales. These practices interrupt hegemonic structures and representations of space, both because they are hard to know and to control, and because they directly challenge existing spatial orders (98).

Similarly, Don Felice reminds us that urban spaces are structured and built to encourage certain uses, to transmit certain ideas, or to make possible certain conversations. As he explained, some areas of Milan are more accessible to new residents of the city and lower-income people than others. Some restaurants do not welcome visible minority patrons. At the same time, however, people's journeys through, uses, and interpretations of spaces interact with these dimensions of space, making the urban terrain unsettled and contested.

Don Felice alluded to this dynamic several times during his interview and walk. The Navigli neighbourhood, he noted, is built around the canal and structured by it. Moreover, most of the people who are there at night seem to come from outside the neighbourhood. This speaks to the fact that much of this area has been gentrified: the CD vendors and many of the people who are "walking around" most probably cannot afford to live there. At the same time, by their being there, people build an improvisational, fluid public space that is only, in part, reducible to the formal and material characteristics of the built spaces.

Similarly, Don Felice described the centre as a space organized by fashion stores and advertisement. He explained:

This public space [...] is constructed by the [...] visible design and symbols of brand[s].

According to Don Felice these images and this retailing, most of which is unaffordable for many people, structures this space by giving it an elitist allure. At the same time, however, the expensive clothing shops cannot keep less desirable people and their wandering bodies away, because, after all, "the space is open [...] everybody moves, back and forth, back and forth." In this context, Don Felice pointed out,

to enter this space is like taking a bath in different brands [immersing oneself in the light of the brands' displays reflected from the stores onto the sidewalks and streets] ... one comes out and says: so, I saw Armani, I saw [that other brand].

One's experience of this place is inevitably shaped by fashion and affluence, but the very movements of people in and out of it contradict some of its consuming logic and make room for other understandings and uses of this space.

According to Don Felice, the way in which immigrants claim and utilize streets and piazzas in Milan is yet another example of the daily reconstruction, reinvention, and appropriation of space by residents of the city. As he argued above, immigrants' use of the Duomo Piazza and of parks render them "open" spaces – locations that are susceptible to negotiation and surprises.

My itinerary in 2009 through the centre of Milan with Riva, a young woman originally from Eritrea, is another striking example for this. Throughout our walk, Riva showed me how an everyday itinerary can be at the same time a difficult journey through a racist city and a subtle practice of becoming a resident. Riva bitterly decried how she is often singled out and discriminated against as a Black woman, and as a nanny. She recounted, for example, the widespread rudeness of children she encounters. Clearly, her feeling of always being "the other" greatly shaped her experiences in Milan. At the same time, she talked about her and other immigrants' active engagement with the positive aspects of the city, including its cosmopolitanism, its economic opportunities, vibrant public spaces, and its beautiful street fashion. Indeed, our walking tour itself became a search for inexpensive stores that offer resources for dressing beautifully and thus becoming more "Milanese" in the eyes of other inhabitants of the city. Most importantly, Riva was not only showing me stores that she liked. Rather, there was a sense that it was only through partaking together in the embodied, performative, everyday practice of *becoming local* – through dressing, window-shopping, and aesthetic appreciation – that she could express and point out the active and constant work of shaping one's subjectivity.

Riva's and Don Felice's observations – and the very enactment of this practice during our walks – suggest that the city is not simply a fixed place to be discovered and used by social actors whose identities are settled and certain. Instead, a walk through the urban terrain has the effect of reframing both places and people. Their key insight is that, because both public spaces and selves result from dynamic processes of creation, using and journeying in city spaces can be a way of reinterpreting and rewriting diversity itself. As a strategy for this line of thinking, Don Felice looks at the city as what each person continuously discovers through journeys, stories, and encounters. In his words, the park picnics of Peruvian-Italian people are as "exotic" as the "baths in the brands" in the centre of the city or as the gathering of people by the

Navigli, the canals that are one of the quintessential markers of Milanese identity.

His question, "How do we live as strangers in the city?" is linked with his observations that one is *made* into a stranger in the city through acts of exclusion and discrimination, through structures fostering inequality, and through economic disparity. Moreover, what I find most significant in his commentary is that Don Felice suggests that we look at being a stranger to place as something that all residents of the city share, albeit in different ways and for different circumstances (including because they leave it every weekend to go to the mountains or seaside). His description of public space as a sand dune is a key idea in this respect. It is because public space is a sand dune, shifting and ephemeral, differently constituted in various locales, unpredictable and replete of encounters, and because it harbours so many stories, memories, journeys, and relations of power, that even people who have lived in Milan all their lives can find themselves strangers to a particular space.

Taking the indeterminacy of public space as a starting point to talk about the construction of difference, Don Felice's commentary reverses the common thinking in Milan that interprets the use of space as a direct result of cultural identity. He suggests that tracing the dynamic, powerful processes that make us "native" and/or "other" might be a more fruitful approach than equating subjectivity with a fixed spatial belonging (Narayan, 1993). His insights are an important countertalk to the strategies adopted by the municipality in recent years. Increasingly, efforts and investments geared towards a meaningful entry and participation of newcomers in Italian society have been substituted by programs seeking to compartmentalize and displace them. A "new racism" centred around cultural difference as an unbridgeable gap (Amin, 2004: 15) legitimizes policies of separation. In this context, Don Felice's questioning of how each inhabitant is a stranger in the city – in crucially different ways and according to often opposite processes – raises more critical attention to how inhabitants move and meet in urban locales.

Chapter Six

Entangled (In)visibilities

Carnival Week occurs every year between mid-February and late March, and it brings large numbers of people to the streets of Milan. It lasts from Tuesday to Saturday, but the last three days are the busiest ones, culminating in the parade on Saturday that runs through the major avenues of the centre of the city. Those who wish to, whether adults or children, can wear costumes. On Saturday, 12 February 2005, the parade filled the Corso di Porta Venezia, San Babila, and ended at the Duomo Piazza, which by the afternoon was so filled with people that it was very difficult to access it. Following the parade as best as I could, and observing the sights, sounds, and movements in the streets around it, I was struck by the multiplicity and reciprocity of gazes in public space.

People in costumes were parading in front of onlookers, as models from a giant billboard were staring at both the Carnival personages and the watching crowd. Moreover, as the carnival masquerade was not limited to the parade, but was diffused in the streets of the centre, the difference between the looking and being seen of Carnival goers and of the people who were window-shopping and doing their weekend promenade was just a matter of degree. The two practices intertwined, mirrored, interrupted, and added to each other in various ways. Carnival figures were showing their costumes not that differently from how they might a nice dress, and shoppers alternated looking at costumes, at regular store displays, and at the people walking in their fine weekend clothes. Just in front of me, three masked persons were gazing into a shoe store, their colourful reflections in the glass. When they entered the store to check out the sales, they could be seen from the outside of the shop, thus becoming an interesting part of the merchandise display. The glass of the windows had subtly become a surface around which Carnival ended and restarted.

Figure 23. Milan's Duomo Cathedral with a giant jeans ad hanging from its side. (Dec. 2004).

Corso Vittorio Emanuele and the Duomo Piazza flooded with people watching, showing, and concealing. These different acts of seeing had different goals, and were directed at different subjects. Layered onto each other, like the parts of a cake, they were generating "modes of attunement, attachment, and composition" (Stewart, 2008: 71). If the people circulating in the centre were using streets and piazzas as stages for creating selves and society, they were doing so through subtle ways of being and seeing in the piazza that were at the same time ways to relate to other city dwellers.

A similar sense of visual saturation pervaded the Carnival in 2009. Indeed, the "web of looks" (Taylor, 1997: 19) during that event was even more complex, because the Carnival days of that year coincided with Fashion Week. Responding to critiques in the media that the fashion events did not offer anything for regular inhabitants, and that the designers had grown disconnected from the general public and from culture at large ("Montenapo il commune accusa gli stilisti," 2009), the

local Fashion Association installed giant screens in several central piazzas, so that people could see the catwalks that were unfolding in the
showrooms.

One of the screens was placed in San Babila. When I passed by
there on the Thursday afternoon, nobody was watching it. In the quiet
bustle of the piazza, everyone was ignoring it, concentrating instead
on their friends or on their cell phone conversations. As the Saturday parade unfolded, however, and people gathered in large masses
in San Babila to see the costumes, the screen and its images became
surrounded by balloons, flying confetti, Carnival carts, and masks.
Slowly the catwalks depicted on the screen seemed to lose their air
of elegance and unaffordable distance. Looking at them from amid
the Carnival crowd, I wondered: was fashion being made ridiculous
by appearing to be just another masquerade among many? Or, on the
contrary, had it managed to remain at the heart of complex practices
surrounding identity, dress, and social relationships in the city (see
Figure 24)?

Adding yet another set of costumes and performances to the crowded
piazza, the screen's catwalks showed both the difference and the sim-
ilarity between fashion and Carnival as they manifest themselves in
public spaces. Carnival seemed to offer an inclusive way to be there,
visually and performatively (at least to a certain extent, everyone was
invited to participate), but expensive fashion in Milan had generally
worked to strengthen social distance and difference. Yet, both Carnival
and fashionable dress relied on showing oneself to the gaze of others.

Interestingly, it was this last feature that made the screen particularly
vulnerable. In a subtle way, more than by the critiques raised in the
newspapers, more than by some of my interlocutors' comments about
it being unaffordable, snobbish, and exploitative, the authority of fashion represented on the screen was being challenged by its audience not
taking it seriously. Engulfed by a different set of visual relations, the
catwalk on the screen had ceased to "work" as it should. The stories it
told were being deconstructed by its very surrounding.

Of course, we could argue that the screen was a way to "pacify" the
common people, while the "real" business was happening elsewhere,
in closed rooms. Who looked at the screen and what they thought about
it was ultimately irrelevant to the millions of euros in sales made during and after the catwalks. In this sense, we can look at fashion as an
economic system driven by forces and processes like ownership and
production, and in which aesthetic practices are a superficial product.

Figure 24. The fashion screen in the San Babila Piazza. The catwalk's screen on Thursday (*top*) and on Saturday (*bottom*) during Fashion/Carnival Week. (Feb. 2009).

However, another interpretation is possible. Several of my interlocutors in Milan suggested that the "struggles occurring at the level of the visual" (Pinney, 2004: 8) and taking place in public spaces are an important part in the way not only fashion, but also social difference operates in the city. As a social praxis taking place in particular urban settings, vision partakes of the same dynamics, unequal relations, conflicts, and dilemmas that characterize public spaces. In turn, the latter are constructed and negotiated also through visual processes and conflicts. Following their insights, what I seek to do here is to explore how looks, visibilities, and invisibilities might "make things work," instead of taking aesthetics as necessarily an illusionary and surface effect of real forces, as "simply a reflection of something else, something more important happening elsewhere" (ibid.). As Christopher Pinney argues, sometimes practices of seeing and images can make history (ibid.), rather than always and inevitably the other way around.

This approach is particularly helpful for understanding how (in)visibilities manifest themselves in Milan and how they help shape, and are shaped by, social relations. For this I turn to the critiques and insights of four groups I met in Milan: Terre di Mezzo, Naga, the Stecca degli Artigiani, and the participants of a workshop called "Building Zenobia."[1] Terre di Mezzo is a street newspaper as well as a publisher of books on Italian and international social issues. Naga is a free medical clinic and support centre for immigrants and refugees; it comprises a "street medicine" team that works at night and offers free health care to homeless persons and to inhabitants of makeshift settlements and/or camps. The Stecca degli Artigiani can be counted as one of the Social Centres: when I first visited it, it was an "occupied" former manufacturing premise, used by several groups for cultural, political, and social projects. "Building Zenobia" was a workshop held at Casa Loca in March 2005; during this ten-day initiative, architecture students, activists, and professors designed possible floor plans for a renewal and use of a dismissed area between the municipalities of Milan and Bresso. Planning possible settlements, "Building Zenobia" participants wanted to widen debates on the use of spaces in Milan.

The critiques of these four groups were illuminating for me in several ways. First, by using a language of vision and visibility in order to denounce discrimination and exclusion in the city, they offered not only a way to think about social processes of marginalization, but also a critical understanding of visuality. Second, for the participants of "Building

Zenobia" and members of the Stecca, attention to invisibilities was a strategy to rethink the role of dismissed areas (*aree dismesse*) in the city and some of the current processes of urban renewal, and to counter gentrification and inequality. They pointed out that tertialization (and the rise of the economy of fashion and appearance, in particular) and the vast expanse of dismissed lands and buildings are the two sides of the same coin. What intrigued me in their analyses, moreover, was their insistence that these two sides need to be linked with each other not only for the rather obvious reason that deindustrialization has been changing the urban territory, but also because the former is a force that reorganizes the visibilities and invisibilities of both spaces and people. This is because fashion and aesthetics – as an economic and cultural system, as a performative practice taking place in urban locales, and as a post-industrial economy allied to urban revitalization projects – affects the social position of inhabitants, thereby shaping the ways they can appear, literally and figuratively, in its landscape. From this point of view, post-industrialization can be analysed also as a form of intervention in the visual field – with the latter understood as shaping and constituting social relations in the city rather than simply representing them.

According to participants at "Building Zenobia," these processes are essential to understanding the changing role of spaces in Milan. Their analyses did not distinguish between public and private spaces, or between public spaces in particular and urban spaces more generally, because what is at stake here are the very decisions and processes that define city spaces along those boundaries. Following these situated lines of reasoning, and adopting them as a theoretical reflection on space and vision, in this chapter I use invisibilities as an approach to talk about space and inequality in a changing city.

This chapter is a response to a particular ethnographic puzzle. My interlocutors emphasized that it is necessary to put in relation ideas and processes that seem at first sight not related to each other, such as the fashion system and the marginalization of many inhabitants, the precariousness of existence, and the spatial haunting of the Milanese territory. Theirs is a complex critique that jars more habitual ways of understanding the city and its inhabitants. Mirroring its associations and juxtapositions, in this chapter I discuss side-by-side practices of invisibilities that marginalize inhabitants and fashion as a cultural, economic, and political force that mediates social relations in Milan.

Seeing, Being, and (Dis)appearing in Milan

Terre di Mezzo (2003) and Naga (2003) describe Milan as a metropolis teeming with "hidden cities." These are composed of homeless people who live in the streets, the train station, and in parked vehicles (Acanfora, 2004: 4–5; Giorgi, 2003: 17), immigrants and refugees without papers who create temporary settlements in empty buildings (Naga, 2005 and 2003), Roma and Sinti families who live in camps and villages in the periphery (Acanfora, 2003: 15), and low-income migrants who subsist through small temporary jobs and the assistance of shelters (Chiari, 2004: 16–17). These two organizations talk about an "invisible," "forgotten," and "underground" city to point out the social exclusion of these and many residents of Milan. With these terms, Terre di Mezzo and Naga point out that the municipality and many inhabitants of the city treat low-income immigrants, the homeless, and Roma and Sinti people as if they were simply not there. Their rights and social realities are ignored, and they are de facto excluded from full citizenship in society.

This invisibility manifests itself differently for various groups of people in different circumstances. Simply distinguishing between people who are visible and those who are not is hardly helpful. Moreover, by equating invisibility with exclusion, we would find people who are totally marginalized and people who are totally powerful. We might thus overlook how powerful interests sometimes work better precisely by being invisible (see the "Building Zenobia" participants below). Vision can be better described, using Gordon's words, as "a complex system of permission and prohibition, of presence and absence, punctuated alternatively by apparitions and hysterical blindness" (1997: 15, referring to Kipnis, 1988). As Gordon makes clear, moreover, invisibility can include both not being seen – for several reasons and processes – and ironically, being way too visible.

This idea can help us understand the contradictory position of some disadvantaged residents of Milan. Homeless people are metaphorically invisible as social and political subjects, yet they are very – indeed, too – visible in urban public spaces (Mitchell, 1995). Their presence there does not correspond nor lead to a participation in society and in the public sphere. In fact, the former negates the latter. In this context, homeless people's hypervisibility becomes a matter of "I see you are not there" (Gordon, 1997: 16).

Similarly, low-income visible minority residents of Milan are at times invisible and at times very noticeable, depending on where, when, and how they occupy piazzas, parks, and streets. A friend and colleague of Don Felice, Don Pietro, explained that immigrants who hold several jobs to make ends meet often do not have any time to use public spaces. Especially if they are employed in small factories or in private homes, these workers remain hidden from mainstream discourse (Caritas/Migrantes, 2005). Because their work keeps them away from urban places of visibility, it does not counter widespread negative perceptions of unemployed, "unproductive" migrants. Their absence from public view is an easy alibi for politicians and municipal structures to do very little to track down abuses at work, or to make housing more accessible.

At times, many immigrants are necessarily very present in public space. As the Schuster Youth discussed, this can result from some groups using public spaces as sites where they can link with co-nationals. It can also be because of an occupation that particularly invests public spaces. This is the case for street vendors and buskers and for some domestic workers. A Latin American–Italian woman I met in the Scala Piazza, for example, described how she usually rests while waiting in streets and piazzas between her shifts in private households of the Centre.

It must be emphasized that in all of these examples, (in)visibility is not simply a metaphor for social processes.[2] The argument I want to make here is that – because in Milan embodied and situated practices of seeing are active ways to participate in public space – visibility, representation, and concealment are powerful social practices that invest material forces and imaginative work in the tight embrace of place-making. The noticeable visibility of people who are perceived as non-Italian translates into alterity, constructing them as intruders and/or out of place in the eyes of many Italian residents (Monteleone and Manzo, 2010; De Giorgi, 2010; Zajczyk, 2005; Maritano, 2002). Because of this, immigrant users of public spaces are often the first to be displaced when it comes to urban renewal (Dines, 2002). Another consequence this can have is that middle- and upper-class Italian-born people might prefer at times to retreat from public spaces, instead of sharing them with publics they do not feel connected to.

In turn, a number of laws, regulations, and policies seek to manage and control migrant bodies by casting them out of view. Paone (2008) and Van Aken (2008), for example, argue that in Italy stowing away the other through camps, transitory centres, and detention sites located at the margins of the city has become the privileged way of welcoming, knowing, naming, and encountering new migrants seen as a "humanity in excess" (Paone, 2008: 85). This follows a wider Europan and global strategy that deploys "spaces of containment" as a way to keep undesirable subjects far from cities and places of work, support, and sociality, making them into "abjects" rather than active social beings (Isin and Rygiel, 2007: 177). According to De Giorgi, in fact, it is estimated that up to a hundred thousand immigrants are held each year in the 218 detention centres located in various parts of Europe (2010: 155).

The Security Package (*pacchetto sicurezza*) discussed in 2008 (and passed, in part, in 2009) by the Berlusconi government can similarly be seen as a tool to perpetuate invisibility. Because of some of the proposed measures, sick immigrants without papers would not show up in hospitals, and children of illegally residing parents would not attend public schools for fear of being reported and/or deported. In Milan, moreover, municipal by-laws introduced by Moratti intensified the control of piazzas and streets, in an attempt to do away with street vendors and panhandlers. Taken all together, these measures reveal the "intent to displace and reject" (Paone, 2008: 110), rather than integrate newcomers in society. Paradoxically, the hypervisibility of migrants and refugees created by labelling, and the intense surveillance by the institutions, is aimed at converting all the spaces of transition wherein immigrants find themselves into places for "removal and invisibility" (105).

The policies of evacuating the settlements of marginalized inhabitants is another dramatic example. Ottavia and Alberto, two members of Naga I interviewed in 2011, explained that the approach of the Moratti government to Roma people's marginalization and precarious living conditions has been limited to further displacing them. Ottavia explained:

> From the municipality the response has been to carry out big evacuations, to dismantle the settlements: to reduce the number [of people], to reduce the number of the camps, and obviously there is no [long-term] solution in this way, because the people continue to live, dispersed in the territory [...]
> This is a matter of image [for the municipality]. There is the evacuation, and then if they come back [it does not matter]. [The deputy mayor] De Corato has counted his 500 evacuations, he has celebrated his 500 evacuations [...] But the politics is like this: is a play of image: being able to say: I have dismantled [the camp]. There is not even the idea, not even the initial approach to say we could think in a different way. (14 June 2011)

These are discourses and practices that dramatically further poorer inhabitants' invisibility. Bonomi (2008) has reported that as a result of the frequent municipal evacuations of their makeshift settlements, in the past few years many refugees, Roma people, and homeless immigrants have been moving from one urban interstice to another. An example of this is the closing of the Barzaghi Camp, where about a thousand refugees from Kosovo and Bosnia were living. Following the clearance of the settlement by the municipality, its inhabitants continued to move between temporary accommodations, including illegally occupied premises, deindustrialized areas, and vacant terrains (200). Ottavia explained:

> The whole politics of evacuations that started with [the deputy mayor] De Corato in the last few years has resulted in people becoming spread out over the territory.
> [...] Now there are only few large settlements, otherwise they are all small groups dispersed here and there. The problem with the continuous evacuations is not just that there are only small groups, but also that the people live in increasingly precarious conditions. One goes from the motorhome to the tent [...] And then there are more and more [people] living in interstices, so that they are not seen.
> Very often they are close to train tracks. While before they were under the bridge – already dramatic situations – following this dispersal you tend to find yourself between a wall and the train track, or on the terrain between train tracks. So you become more and more invisible, and are in more and more dangerous situations. (14 June 2011)

All of the dramatic inequalities and struggles that I have described – and which are expressed or carried out in and through the visual realm – take place in a social, cultural, and economic context in which aesthetics and visuality always already play a part in how people live in and talk about their city and relate to other inhabitants. In particular, from the 1970s onward, fashion became not only an important industry in Milan but also a social idiom, a cultural system with crucial links to politics and culture (Segre Reinach, 2005; Foot, 2001; Bourdieu, 1984). This is a complex conjuncture that requires us to pay attention to very different aspects of the social and of Milanese visual engagements at the same time.

In the 1980s, especially, Milan was the epicentre of the prêt-a-porter fashion sector associated with names such as Armani, Krizia, Laura Biagiotti, and Missone. Whereas since the 1950s Florence and Rome had been the most notable centres for clothing, promoting high-end and boutique fashion through their catwalks, starting in the 1970s many designers chose to showcase their work in Milan (Gnoli, 2005). Interestingly, this phase of the emergence of prêt-a-porter fashion was intimately linked with the already established field of design and its functional aesthetic and, more generally, with "a sense of cosmopolitanism" (Segre Reinach, 2006: 124) created by people involved in the cultural field including journalists, artists, photographers, and designers.

As fashion became more and more successful and took on the role of economic motor in the city, fashionable clothing and accessories became tightly enmeshed with urban locales. Clothing stores took over the centre of Milan, displacing many traditional stores that had existed for years. Many piazzas became stages for catwalks, and the central *quadrilatero della moda* (Quadrilateral of Fashion) became an unwelcoming neighbourhood, where only the very wealthy could afford to shop. This intensified and hastened gentrification in central areas of the city. Moreover, one of the most significant redevelopment projects currently being realized, the Garibaldi Repubblica area (discussed in more detail below), includes the construction of a new "Fashion City."

Fashion then dresses not only people but, in a sense, the city itself. Advertising, giant screens, and the sponsorships of green spaces are just some of the things that make adornment and particular notions of style a recurrent aspect of people's experiences of public spaces. In addition, because most shop windows start only a few inches from the ground, often the sidewalk seems a continuation of the store and vice versa. Because the rise of Milan as a city of design and fashion increased its self-ascribed sense of cosmopolitan sophistication, many Milanese I talked to during my research told me that the *struscio* (the promenade) does not exist anymore as it is too provincial and antiquated a habit. Ironically, however, the same processes that have made the struscio appear less central have introduced practices that displace it to other contexts or borrow some of its key characteristics.

The mega-ads are an interesting example for this. A person working in this industry, who I interviewed in 2009, explained to me that originally these ads were giant paintings, realized by the same people who were producing theatre sets. Although this technique has changed in the past twenty years, the mega-ads still have to be understood as scenographies. Mirroring and framing what goes on in the city, they act as stages for struscio-like engagements with urban spaces, which however are more generalized and diffused and, increasingly, include advertisements and shop windows as part of the "mutuality [...] of vision" (Pinney,

2002: 355). Store windows, new marketing techniques, and the model of the café-boutique employed by Gucci (see "Orientations") all similarly use the theatricality that has always been so central to the struscio to create relations between the store and the street, and the consumers and the brands. When we take all of this into consideration, we can say that the processes of deindustrialization that replaced factories with the production of style, fashion, and "beauty" have sustained and possibly expanded the persistent role of vision in social encounters. At the same time – because, in this process, economic and social relations in the city have hardly become more equal or just – the question of who is welcome to participate in these dynamic landscapes, and how, is more and more pressing.

According to Simona Segre Reinach, all of these developments have concurred with an "aesthetization of every aspect of everyday life" (2005: 81), in which politics, class, gender, and diversity are increasingly debated and perceived in aesthetic frameworks. This is partly a result of the democratization of fashion brought on by prêt-a-porter. High fashion was primarily an "instrument [...] of distinction" (ibid.; see also Bourdieu, 1984) and was centred on the concept of class, but prêt-a-porter designers integrated street styles and youth subcultures in their products and usually offered multiple lines, which included more exclusive collections, as well as more affordable and mass-marketed versions of the same brand. Mirroring and articulating changing notions of class and social differences – increasingly understood as a matter of different lifestyles (see Segre Reinach, 2005: 82) – this new fashion became primarily "an instrument of communication" (81) that helped people create and express their identities.

Significantly, Segre Reinach argues that it was a democratization of fashion not so much because its products were more accessible economically, but rather "because fashion becomes a cultural need" (2005: 89), a required tool for participating in society. Today, writes Segre Reinach, not only is "aesthetic sensibility [...] an essential and integral component in the everyday life of Italians" (1999: 32). Aesthetics has become a way to understand and frame a variety of experiences and realms, including politics, science, technology, and morality (ibid.).

My personal experience, of attending high school in Milan in the 1980s, the golden time of prêt-a-porter, echoes Segre Reinach's observations. Our class was deeply divided along political affiliations, yet none of us students ever talked in explicit political terms. Instead, very clearly defined dress codes reflected a left-leaning commitment to civil

society, social equality, and multiculturalism, on the one side, and conservative capitalism, on the other. On the contrary, a dear friend a generation before me, who went to high school in the 1970s, talked about
the left, about Marxism, and about workers. I remember being both
fascinated by his knowledge and deeply disappointed in realizing that
that language was not accessible to me. Here I do not wish to general-
ize my experience to a whole era or all high schools, nor do I intend to
say that the language of class and the political concepts of left and right
have ceased to be important. I did eventually learn to speak also that
way. And in the Social Centres people talk about comrades, the oppres-
sion by the ruling classes, and the internationalization of capitalism. But
I do want to point out how the "diffuse aesthetization of experience"
(Ferraris et al., 1995, quoted in Segre Reinach, 1999: 32) helped fashion,
style, beauty, and aesthetics become an important social idiom.

To appreciate the staggering complexity of the relationship between
fashion, culture, politics, and city spaces, it is important to remember
that the rise and the reign of prêt-a-porter coincided with a period of
great crisis for the city, the "Years of Lead" (*Anni di Piombo*, or "Bullet
Years")." In this context, according to Segre Reinach and to Gnoli (2005),
fashion has served a double function. For one, as a dynamic cultural
form and social idiom available (although differently) to much of the
population, fashion managed to bridge some key tensions and ruptures
of that time such as the contradictions between "the oppositional values
of the seventies" (Segre Reinach, 2005: 104) with its critique of bourgeois emptiness, on the one hand, and the focus on beauty, individuality, and "the care of the self" (ibid.), on the other. Simply put, fashion
and aesthetics took the role of a shared conversation between widely
different speakers. For another, fashion served to dilute, and in a way,
helped one "forget" the political tensions of that time, without contributing possible solutions. In fact, because high-end fashion is expensive
and not everyone has access to it, inevitably it reinforced disparities that
were at the very basis of those struggles.

Finally, the role of fashion in Milanese public space does not only
include people's engagements with fashion and style and aesthetic cultural norms. The tertialization that transformed Milan (and of which the
fashion system is a part) (re)shaped neighbourhoods and public spaces.
In this process, it affected the actual and metaphorical visibility of particular subjects and locales and sparked, in turn, tactics of resistance that
are carried out and expressed in a visual idiom. Post-industrial empty
terrains, for example, are sites where differently positioned social actors

construct and deploy different kinds of visibilities and processes of concealment.

Ghosts

> What kind of case is a case of the ghost? It is a case of haunting, a story about what happens when we admit the ghost – that special instance of the merging of the visible and the invisible, the dead and the living, the past and the present – into the making of worldly relations and into the making of our accounts of the world. It is a case of the difference it makes to start with the marginal, with what we normally exclude or banish, or, more commonly, with what we never even notice. In Gayatri Spivak's formulation, it is a case of "what … it [is] to learn, these lessons, otherwise."
>
> – Avery Gordon, quoting Gayatri Spivak

If vision as a tactile, mutual, and relational practice serves to negotiate identities in public spaces, if the way the city looks helps frame debates and understandings, Avery Gordon (1997) also reminds us that the social is always haunted by spectres. These emerge at uneasy places of encounter between different positions, histories, and interests, and bend common ways of seeing, appearing, and concealing, as they show subjects and relationships that are supposedly not there, or given-to-be-invisible (see also Taylor, 1997). Gordon's ideas echo the insights of some of the people I met in Milan, who taught me to see ghosts in the Milanese landscape. Their critique is particularly helpful in understanding the impacts of the vast abandoned or dismissed areas in Milan. Rather than conceptualizing them as empty lots and buildings, my interlocutors argued that they are places where invisibility as a social and political condition becomes particularly manifest.

This is the case, for example, for many illegal immigrants who are residing in these areas. Naga argues that many illegal immigrants are doubly invisible subjects, because they have to render themselves so. Their social marginality both perpetuates and requires them not to be seen and noticed. Their lot is made even more dramatic by the fact that they inhabit invisible places. Many immigrants without papers, in fact, occupy empty industrial premises in dismissed areas or similarly abandoned sites (see also Multiplicity.lab, 2007). As a Naga publication states, these accommodations are generally without water, electricity, heating, and other services. The situation is mirrored by the lack of engagement from the part of the municipality, who according to Naga, defines them as "non-persons" (echoing Dal Lago, 1999) and reasons that "as they are not regular [i.e., are illegal] residents in Italy, they should not be there, and thus they are not there" (Naga, 2003: 36). Naga points out the stark contrast between the "immense spaces" (6) of these vacant buildings, where groups of usually a hundred to three hundred people try to create a home, and the necessity of not being seen – a need dictated by the illegality of the residence, by the lack of papers, and by the disruptive incursions of the police or, even worse, of neofascist groups. Indeed, the very possibility to live in hiding in such a huge place derives from the peculiar status of these areas that work like gigantic "black holes" (Zenobia presentation, 9 March 2005) in the city terrain.

The participants of the workshop "Building Zenobia" explained that illegal immigrants who are the users of dismissed areas are not the only "invisible subjects" linked with these spaces. The rightful owners of these properties are also hidden, because this is advantageous for their productive operations. One participant explained that, in Milan, dismissed areas

say a lot about global conflicts, that is, of conflicts between invisibles. Because in the end, in the dismissed areas in Milan, there are two things happening: there are possessors [i.e., users] and owners. The possessors are invisible because they are migrants, mostly illegal, and thus have to make themselves seen as little as possible. The owners, on the other hand, are invisible because they are the big owners, transnational ones (because there are some that are Italians, but many that are international), whose goal is to make themselves seen as little as possible, because so they can work better. So these conflicts in the dismissed area – closed, invisible area, visible [only] from up high […] – are invisible to everyone because the subjects [of these conflicts] render themselves invisible because of their different needs.

The particular area that the participants of Zenobia focused their workshop and projects on is a perfect example. It is a former factory between the municipalities of Milan and of Bresso, and it has been abandoned for years not because it is unusable or unnecessary, but in a sense because it is too valuable and thus the object of different and conflicting interests. It is owned by Auchan, a large private distribution company seeking to build a shopping mall on it. The municipality of Bresso, however, opposes this idea, as a big mall would disrupt the small retailing that Bresso has been supporting and promoting within its boundaries. The fact that a portion of this area is located outside of Bresso, in the municipality of Milan, renders things even more complicated (see Caronia, 2005).

This is why participants in the "Zenobia Workshop" contested the very word "dismissed area." Veritable dismissed areas, they explained, are spaces that cannot be used because they are too polluted and/or too costly to fix. Instead, dismissed areas in Milan are political "back holes" that are traversed by "a whole mass of interwoven interests" (Zenobia presentation, 9 March 2005), both private and public, that prevent them from taking shape.

Zenobia's focus on the (in)visibilities of places, subjects, and interests is illuminating in more than one respect. First, it draws attention to the politics of urban planning and renewal in Milan, and how they respond more to the market advantages of the few rather than the needs of the many. As one of the participants pointed out:

> Milan is a city filled with families who need [housing], on one side, and full of empty spaces, on the other.

Second, it suggests that to understand some of the things that are going on in contemporary Milan it might be useful to examine the invisible links that sometimes exist between social subjects in the city. According to Zenobia, these links do their work precisely because they are so difficult to see. Third, because of the invisible characters and hidden interests inhabiting them, the dismissed areas can be thought of as "spatial ghosts" in the urban landscape. This latter idea was a particularly helpful guide during my visits to the Isola neighbourhood, a working-class area close to the city centre that has been undergoing significant redevelopment.

During one of our city walks together, Francesca, a middle-class, Italian-born woman, brought me to this part of the city. Indicating a vacant, grassy space, interspersed with patches of dry earth, shrubs, and sections of fences, and pointing to new developments already starting to take place in the area beyond the field, she said:

> This is [the] Isola [neighbourhood], this corner here. It is full of small shops, cafés, pizza restaurants, night clubs. (15 March 2005)

As Francesca's words were directed at an empty lot, it is hard to know whether she was really just talking about the existing businesses and places in the neighbourhood (some of them visible in the distance, behind the large field) or whether part of her description also included some of the things she imagined would spring up in this immediate area. Francesca, in fact, continued by describing how the whole neighbourhood is being restructured, and new residences built. She emphasized that this is a historic part of the city, one of the oldest.

Francesca's words stuck in my mind for a very long time, as for me they demonstrated, better than any other description, the way in which areas of the city change as they are imagined differently (Potuoğlu-Cook, 2006; Zukin, 1995) and how they acquire different images and interpretations as new inhabitants, urban projects, parks, piazzas, and buildings move in. The fact that to me this still was an empty weed-filled expanse of land, and not a new neighbourhood or an old historic quarter of the city, emphasized the different interpretations and versions these imaginaries might hold for different inhabitants and passers-by (Herzfeld, 2009).

Francesca's ability to see urban renewal in an empty lot is strikingly similar to the strategy of resistance and opposition employed by the organizations of the Stecca degli Artigiani, an occupied dismissed building in the Isola neighbourhood, to counter the gentrification of this very area, in which they have been residing for years. Both Francesca and the activists of the Stecca, then, suggest a way to look at the empty spaces of the city as a possible starting place for analysis, similarly to how Derrida (1976) saw the supplement as holding a central place in reference to that which it supposedly merely adds to. The map of the Isola, created by the organizations of the Stecca and displayed in the neighbourhood park, is interesting in this respect.

As Figure 25 demonstrates, the Isola has long been a working-class neighbourhood. One of its significant places has been the Stecca degli Artigiani, a factory that was part of Siemens-Tecnomasio Brown Boveri, but then was abandoned. Since the 1980s the Stecca has been used by several people and groups as a place for exhibitions and film screenings, a meeting place for grassroots oppositional associations, a political party headquarters, an occupied residence for a small community of immigrants, an association promoting cycling, and a library – just to mention some of its uses. The park in front of the Stecca itself has an interesting history, as it was created by the residents of the neighbourhood who wanted a green space instead of a new parking lot as had been planned by the municipality (see www.lastecca.it and www.cantierisola.org; accessed May 2006).

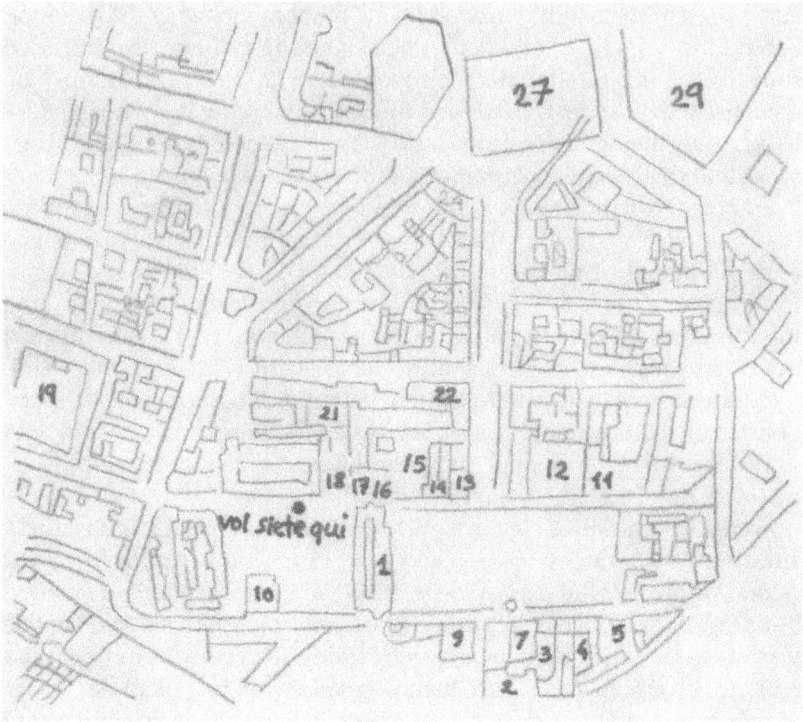

Figure 25. Map of the Isola neighbourhood, redrawn from a sign posted at the entrance of the park beside the Stecca. This map reminds passers-by that "you are here" (in Italian: *"voi siete qui"* – as is written just by the building (1) and the area representing the park), in a place teeming with life, social struggles, and histories, and not just in front of an empty green space that many want transformed into new developments. The structure number 1 is the Stecca degli Artigiani, and the other numbers around it refer to places that were once small manufacturing and industrial businesses (such as 2, a comb factory; 3, a soap factory; 4, an industrial depot; and 5, an ironworker), to a school (19) and a church (21) constructed at the beginning of the twentieth century, as well as to old residential buildings (7, 9–18, 22, and 29), many of them constructed at the end of the nineteenth century; 27 is the site of an old cemetery. (March 2005).

At the time of my walk with Francesca, in 2005, however, this part of the Isola was the site of a planned new development, which is currently converting this area of the city into a new residential neighbourhood and a central node for services associated with fashion. In 2009, in fact, when I returned to Milan, I found it partially replaced by one of the largest work-in-progress zones I had ever experienced. In the Isola and in its wider area, the Garibaldi Repubblica, three connected projects were transforming approximately 290,000 square metres of empty areas (Armentano and Lupatini, 2007), vacant lots resulting from the move of a train station and from Second World War bombing and dismissed buildings previously used for manufacturing. They were generating a new neighbourhood that will be called Porta Nuova, housing (including new residences and renovated older buildings), spaces for businesses, a hotel, the new headquarters for the government of the region and the municipality, a post-secondary campus for fashion and design, a fashion museum, and new public spaces, including new streets, piazzas, and a large park.

These changes were not only significant in terms of the quantity of land and the funds involved: this was the only redevelopment of this scale that was occurring so close to the city centre, and thus giving a completely new look and feel to the core of Milan. The transformation of this area was therefore a profound transition for the city as a whole. To appreciate this sense of epochal change we have to consider that the new site of the Lombardy Region is taking on the role previously played by the landmark Pirelli high-rise (designed by Gio Ponti and constructed in 1955–58). This is not simply a competition between two very high buildings. The Pirelli tower is associated with the years of the "miracle": it represented modernism and the period of economic and industrial development. When on 10 May 2009 the new high-rise became the tallest tower in Milan, it signalled the conclusion of an era and the beginning of another. One of my interviewees who attended an event organized by the municipality to mark the day the new high-rise surpassed the older one in height talked about a "feeling of future" and about history being made in front of her eyes.

Although most people in Milan would agree that the area needed to be redeveloped in some way, there had been no agreement on what this should entail. Since 2001 the associations living in the Stecca had been opposing the changes I describe above on the ground that it would displace many of its lower-income residents, devalue the history of the area, and disrupt the countercultural activities hosted in the neighbourhood. The organizations living in the Stecca were particularly opposed to the demolition of the building, which finally occurred in 2007, to be replaced by residential and office high-rises. As a way to gather support for this space and for the work of its inhabitants, by 2005 the park in front of the Stecca had been hosting for three years a monthly organic farmers' market, which involved both the park and the Stecca itself. Here is an excerpt of my fieldnotes from a market day:

I pass the green gates of the park, and I am greeted by a multitude of small stands. Some people are walking from table to table. Many more are sitting on the grass, enjoying the sun, the company of friends, and the relaxed atmosphere of the place. Some people are strumming guitars, others walking their bikes. Most of them seem to be under 30, although older people are present as well. While most people are white, there are also some visible minority Milanese. Most of the visitors are dressed in Milan's counterfashion: colourful Indian scarves, woven cloth bags, and a general lack of mainstream designer clothes. Leaving the farmers' market, I reach the Stecca, and enter the courtyard of the old building. It is long and narrow, and it overflows with a river of people, walking, talking, and pursuing different activities. A young woman has set up a little shop with hand-made jewellery, and up the stairs, into the building, a clothing swap is taking place. The posters on the wall, and the flyers that are lying on tables and chairs, inform the visitors of the offerings of the day: a children's concert and activities, a public forum (with John Foot), and film screenings on gentrification and urban renewal, a biking tour through the neighbourhood, a group game on critical consumption, and a photographic exhibition of the Isola area. I notice that this is one of the few places where a multicultural crowd is coming together. In the café, a large room that opens to the courtyard, and spills chairs and tables into it, several men and women are offering international foods, cooked by North African and South American groups involved with the Centre. In front of the café, a middle-aged man stirs a giant pot of polenta (a corn dish) next to another cook who is preparing small pizzas. The first man asks me, "Do you want something to eat?" and immerses his hands in a large bag of flour at his side, laughing loudly. It is a feast, a celebration, a day of resistance and intense sociality. The market calls people to the Stecca and the Stecca flows towards the market. (13 March 2005)

Here what interests me particularly is the strategy of opposition employed by the Stecca to urban redevelopment projects. As the exuberant gathering and rich program of the Stecca on the market day attest, and as the map that is Figure 25 demonstrates, the activists of the area especially challenged the notion that this neighbourhood was largely made up of "empty" land. This definition, according to them, was a strategy by the developers to pre-empty the area of the many links, associations, and sociocultural meanings associated with it. The market and the events hosted by the Stecca were a way to showcase these networks.

Similarly to the participants of Zenobia, however, rather than simply proposing that these areas are not empty but "full" – for example, full of memory and of users – they suggest that we give a different meaning to emptiness itself. One of the many activities carried out during the market was a so-called tour of the void. This tour included, among others, the Stecca, an abandoned comb factory, and "illegal gardens" on vacant lands (from the flyer of the event, 8 May 2005[3]). By showing the empty lots, which are only imaginatively filled with new developments, as well as the buildings that exist and are inhabited but many investors already see as demolished and replaced, the activists sought to point out the discrepancy between what is (not) there and what many imagine and plan there (not) to be. This, of course, holds true for both sides, as both activist tactics and dreams and future development plans are, in part, solely imagined and often not yet there. It is important here to note that performative practices – the market displays, the festival-like atmosphere of the Stecca, the "tours of the void" – are, once again, key elements of this reflection on and redefinition of spaces.

A contemplation of the void has analytical significance here, because it suggests that the gaping holes in the city are not dead but rather productive places, sites where meanings and things are created, both for real estate developers and for oppositional groups[4] (Goldstein, 2010; Weszkalnys, 2010). This explains, once again, their peculiar (in)visibility. Milanese dismissed areas are akin to ghosts, because they are invisible to many inhabitants, and in many contexts, yet they are starkly noticeable in their many apparitions. Moreover, it is by their very presence at the edge of the dead and the living that they have profound effects on society (Gordon, 1997). As ghostly spaces, they return from death as unsettled terrains. Their demise as factories makes them into very precious grounds for new development, yet this very abandonment as industrial sites haunts them and the city, as they meddle with urban plans, regulations, ownerships, claims, and conflicting goals.

According to my interlocutors, these areas – that are productive sites *because* void – afford unique perspectives on both the post-industrial city of fashion and the way it privileges particular subjects, and on invisibility as a category that ought to be examined, precisely because rendered so by particular social forces. The dismissed areas, the economy and politics of fashion, and the marginalization of undesirable inhabitants are all different yet deeply intertwined aspects of the neoliberal city. They engender practices of vision, concealment, and apparitions that help negotiate people's places in Milan. As such, visibility and invisibility are not external to systems and processes, but intimately enmeshed in the very creation of the social. For this reason, the activists of Casa Loca and of the Stecca encourage us to trace some of the emergent connections between vision as an everyday practice of engagement with people and locales and vision as a part of extensive economic and social urban transformations.

This critique is rooted in, and indeed made possible by a particular – and self-consciously oppositional – understanding and enactment of public space as a medium and goal for political action. It is not simply that for the movement of Milanese Social Centres, some dismissed areas act as liberated, oppositional public spaces (by being "occupied," and thus becoming "liberated" spaces for political action). One of the very interesting aspects of these spaces is that they meddle with and interrogate the very definition of what is, or should be, public and private. One of the reasons that dismissed areas make perfect public spaces, according to my interlocutors of the Social Centres, is because their position as invisible places and voids enables these kinds of questions.

I want to conclude here by returning to San Precario, because as a social ghost he is strikingly akin, and indeed connected to, the dismissed areas as spatial hauntings. Just like the latter, moreover, San Precario is the source of tactics of resistance inspired by the role and power of differential visibilities in Milan's spaces. San Precario is what I would call an apparitional figure: one whose power of representation and commentary relies in great part on the way he can alternate being present and being absent. One of the aspects that I find most interesting in this personage, in fact, is how he spans differently visible realms. San Precario is most times invisible in actual streets and piazzas, yet easy enough to find in the Internet.[5] This is not just by chance. It is rather a tactic and a strength, as his shifting (im)materiality and (in)visibility is what makes San Precario so effective as a tool of critique in urban domains. Like more conventional saints, San Precario is powerful because he lives and expands into the heavens of spirits and hyperconnections. His borrowing from Catholicism, a very powerful force in Italian society, and the peculiar (in)visibility it engenders, is what makes his sudden apparitions and performances in contested urban locales so striking and noticeable (first and foremost on Sarca Avenue, which literally and symbolically unites and distinguishes two sides of the neighbourhood in transition), and thus effective as social critique.

As an apparitional figure, a ghost in the capitalist system (see also Comaroff and Comaroff, 2003), and a haunting labourer in a city that is supposed to have no unemployment and the highest per capita income in the country, San Precario, just like the dismissed areas, teaches us that sometimes only what is ghostly or partially invisible can shed light on our social world. It is no surprise then, that the alter ego of San Precario is Serpica Naro, a designer invented by a group of fashion workers to draw attention to the precarious labour conditions in this field. The association that calls itself Chainworkers spent months creating a brand, a website, and media releases, so that Serpica Naro (an anagram of San Precario) could participate in the Fashion Week of spring 2005 and its official catwalk. This was a serious blow for the organizers of the prestigious event. Once the trick was discovered, it was already too late: Serpica Naro was already "wearing social critique" on the catwalk, performing a trenchant satire on the fashion system, the politics of appearance, gender, and labour relations in contemporary Milan.

Its video "pregnant lady" (which was posted on the Chainworkers website but had been performed on the official catwalk for the February 2005 Fashion Week), for example, features a young woman trying to hide her pregnancy – "a dirty secret just like that illegal little letter [a dismissal letter] which you had to sign [before being hired]" – from her boss through a "hidden belly-band" (www.chainworkers.it; accessed Sept. 2007). In another video, clothing items and accessories help a young woman slip effortlessly from one poorly paid, irregular part-time job to another, all while maintaining a façade of beauty, femininity, and cheerfulness.

Serpica Naro casts clothing, accessories, brands, and styles as tools for a social performance that, in the end, is impossible or unsustainable for most women. More generally, Serpica Naro and other activists indicate that the whole fashion industry in Milan, and the wider post-industrial economic and social structure it belongs to, can be seen as a performance that only some people can afford to play. The practices of seeing, looking, and showing it engenders have tragic consequences for the lives of many people and, indeed, contribute to make many residents of Milan and many of its spaces invisible or void. The economy of fashion, for example, depends on workers "without regular hours, without regular contracts, without paid overtime. Without careers and without a future" (Devoti di San Precario, 2004). Rendering some bodies very visible in city locales, it causes others' lives, histories, and dreams to slip into oblivion. Foot, for example, writes in this respect: "Milan today is a dynamic, glittering fashion capital which hides the dark side of the urban dream. The billions of lire [sic] that circulate around fashion shows, design weeks, advertising companies and private television are underpinned by immigrants working the 'dirty' jobs which feed this economy. These immigrants are often 'non-people,' ignored by the political system [...], marginalized within the urban fabric, lacking in economic and political rights. In the kitchens, sweatshops, bars, and building-sites of Milan, these immigrants provide the labour that maintains Milan's extraordinary post-industrial economy" (2001: 181). At the same time, however, Serpica Naro/San Precario, and the engagements with dismissed areas and invisible subjects, constitute a way "to read the city" differently, and to invite people to "open the door onto" these places and use them as grounds for public debate on Milan. The critical interventions of the invisible can provide, for at least the fleeting instant before it is appropriated, a way to tell a different story and to show a different picture.

Chapter Seven

Walking with Women: Vision and Gender in the City[1]

As I was walking in the centre of Milan with Don Felice, one day in the spring of 2005, he told me, "It matters less the destination, than knowing who you are walking with." During my research in 2004–05, 2009, and 2011, I often remembered his words. When I was following each of my different guides through Milan, it mattered less where we were directed than how the social positions, identity, and life stories[2] of my walking companions interpellated particular aspects of the city and used them to establish relationships with the people and places around them. This was the case for two women I met at the end of 2004, who guided me through a series of walks through the piazzas and streets of the centre: Maria Anacleta, who migrated to Italy from the Philippines, and Francesca, a middle-class, Italian-born woman.

During our walk Maria Anacleta asked me to take pictures of her in particular places, and she looked for passers-by to take snapshots of both of us in front of monuments and in cafés. While these pictures depict Maria Anacleta as part of the city, they also confirm her distance from her family in the Philippines and thus became signs of a difficult journey. Francesca, in her tour, actively sought the beauty of streets, piazzas, buildings, and window displays as a way to relate to her surroundings. Her insights on the "look" of Milan were, in fact, what prompted me to learn more about the ways that fashion and design affect public space. At the same time, this very engagement with aesthetics became a way to reflect on her identity and on other people's positions in the city.

In reflecting on these itineraries, my goal in this chapter is twofold. First, I am interested in how vision serves to negotiate spaces of belonging in the city. By following Maria Anacleta and Francesca through streets and piazzas, I trace how they use ways of recognizing, representing, and unveiling in order to participate in Milanese public spaces. Gender emerges as an important dimension to these dynamics: being and seeing in Milan is deeply affected by what it means to be a woman. The journeys of my guides, however, remind us that while gender matters in Milan, it always does so in class- and race-specific ways (Potuoğlu-Cook, 2006; Preston and Ustundag, 2005; Guano, 2003; Razack, 2000). Their walks, then, show some of the contours of gender, race, and class that affect both city spaces and the specific ways in which different women relate to urban locales.

Second, I pay attention to what I have called the heteroglossia of vision – that is, how modalities of seeing, viewpoints, and differently authored landscapes interrupt, intersect, and/or echo each other. Let me explain. At first sight, the two women's itineraries that I describe in this chapter seem unrelated. Maria Anacleta pointed out the presence of Filipino-Italian people and the places that enable her to connect with them. Francesca talked about style, fashion retailing, and artistic heritage. This very difference, however, can be deceiving, because it resonates with wider discourses in Milan about immigrants being in and using public spaces in ways that are *essentially* different from how "Italians" use them. In this context, as Emanuela Guano points out, an attention to aesthetics, beauty, and art can become a "privileged language" (2003: 365) between people who see themselves as belonging to the city. Identifying a landscape and a way of seeing as specifically Milanese, it creates "viewpoints" for "other" people "from which [...] they can observe – and possibly consent to – the race- and class-specific qualities" (359) of that landscape. For Francesca, seeing and enjoying urban "beauty" serves to claim a privileged connection to the city and a sense of being truly "Milanese." Her itinerary echoes hegemonic discourses linking Milanese culture, art, history, and style with Milanese authenticity. Maria Anacleta's itinerary, just like the commentaries of Don Felice (chapter 5) and Mohamed Ba (chapter 8), contradicts some of these imaginaries, by eluding clear-cut articulations of identities in urban space. This suggests that different ways of seeing and being in Milan by different people have to be put in relation to one another, because they all participate in negotiating who can be part of which spaces in the city and how.

Maria Anacleta

I met Maria Anacleta in the Scala Piazza in October 2004. I had been sitting on a bench, observing the coming and going of people and taking notes in my field diary, when Maria Anacleta approached me to ask me what I was doing. This initial encounter led to several other conversations (in 2004, 2005, and 2009) and to three city walks. During one of these tours, Maria Anacleta showed me her home and dictated the following text to me to describe her arrival and her life in Milan (she dictated the text in English, and the emphasized words refer to Italian terms she used):

> I am a Filipina, I am [Maria Anacleta]. I came to Italy [...] with two of my friends [...] I saw Rome, France, the Eiffel Tower. My brother met me in Rome. Then I visited my Mom. My Mom was here in *Milano* [Milan]. I saw her. I have been in the house with my mother and brother for seven months.
>
> When I found a job, I worked in __ [a city a few hours south of Milan]. My employer in __ died, but my *soggiorno* [work and residence permit] was ready. I met many Filipino people here and when I have no job, I work as a *parrucchiere* [hairdresser]: I cut their hair, and manicure them, to earn money.
>
> Now, after three, four years, I am very lonely, I remember my family, I want to return, but I have to wait for the renewal of my *soggiorno*. I cannot go home without my *soggiorno* because without it I cannot come back anymore.
>
> [...]
>
> When I was in the Philippines, cutting hair was really my job. That was what I did. And I made my children study [...] One of them is a nurse, one studied in the hotel business, and one is in computers [...] My husband worked in Saudi Arabia for five years. I am in the Philippines, I am in my shop, I am cutting hair, together with my children. They are still very young. He worked in an oil factory in maintenance, as a power plant operator. When he finished, he came to the Philippines and I told him, "Ok, you are finished working, so I will be the one to work, I will be the one to go abroad because I haven't been." (17 Feb. 2005)

People originally from the Philippines are the largest immigrant presence in Milan (Caritas/Migrantes, 2008), and they represent a very important component of Asian immigration arriving in Italy today (Cologna, 2003a: 45). Until recently, most of the Filipino nationals in Milan were temporary, older migrants, and predominantly female (ibid.). In the past few years, however, family reunions and the percentage of children and youth have increased significantly (Caritas/Migrantes, 2008).

Several authors have emphasized that gender is a very important factor shaping Filipino migrations at both ends of their journeys. Zontini

(2004) and Cologna point out that women in the Philippines are the ones who support the family, both emotionally and materially. They are "the real pillars of life of the community" (Cologna, 2003a: 45). Emigrating and sending remittances home is considered an extension of care for elderly parents, children, and other family members. Women who do not have children often send remittances for nephews and nieces (Zontini, 2004). Maria Anacleta, for example, has been saving money not only for her children, but also for her grandchildren.

In Italy, the availability of jobs in personal services (household work and caregiving) and the fact that women are still seen as the ones who should "naturally" perform reproductive work (cf. Anderson, 1999: 78–9) are further factors encouraging female immigration. In Milan, Filipina-Italian women are mostly employed as domestic workers, nannies, and caregivers of elderly people. Zontini describes this as the "'international transfer of care-taking' whereby the demanding and socially devalued caring tasks are passed on to poorer and more vulnerable women" (Zontini, 2004: 1133, quoting Parreñas, 2001). This both results in a racialization of care and fails to challenge patriarchal relations in the receiving countries (Parreñas, 2001).

Filipina-Italian women, in fact, are taking on the caregiving duties of Italian-born middle- and upper-class women, who are pressed for time and energy by their double and triple shifts. The latter are often in paid employment while retaining almost sole responsibility for child care and for housework. Ironically, then, in Italy and southern Europe, "non-citizens" are central in "sustaining the European family as a viable social, economic and reproductive unit" (Anderson, 1999: 117; Merrill, 2011). Caring for elderly people is a case in point. As the population in northern Italy is aging, many women find themselves caring for both their children and their aging parents. As this development has not sparked any significant support measures by public institutions, many Italian-born, middle- and upper-class women choose to employ a migrant woman to fill in the gap (Muehlebach, 2012; Merrill, 2006). Maria Anacleta explained:

> When I first came here, I worked with *Signora* [Ms] __ in __ [a town on the seaside]. It was only a summer job, for three months. She spoke English, was 84 years old, and had no husband. She had two daughters. It was a very nice place. They had their own big house, with a swimming pool. I cared for the plants, cleaned the surroundings, helped her *cucinero* cook in the kitchen. I went with her when she went somewhere. On Sunday, my

day off, I went to the *parco* [park] and saw many many Filipino people. It is near the beach, and I saw that many Filipino people are working there.

We came back to *Milano* and I found another job. I worked with a woman who was paralysed. I accompanied her. Her husband was a Sicilian, a very good man. I was taking care of the woman. One day the husband was cleaning a chandelier, very up high and he fell. I was very scared, I did not know what to do. I called the ambulance and I carried the man because the *scala* [ladder] fell on him. He went to the hospital and the woman was crying. They asked me what happened. I said I was in the kitchen when it happened. It was an accident. When he went to the hospital, they discovered that he had a *tumore* [tumor], and he died. After that, I took care of the *Signora* [...]

After a while, they changed me because I did not speak Italian very well. They changed me for a Peruvian or Ecuadorian. The daughter told me: "Maria Anacleta, I am very sorry. It is because you do not speak Italian well. My mother is alone now, she has no husband any more." So I went to my mother and told her, "I have no job anymore because the daughter does not like that *io non parlo bene*, that I do not speak well."

When I had no job I looked again for another job. I found another job and worked there, for *Signor* A. But again he got sick and in the month of August he died. I was not there. He died with his family. I was crying. I had no job again! They called me and said, "My father died in *montagna* [the mountains where he was spending the holidays], you can come and get your things."

My mother at the time was still alive. She died that year. It is very very hard, if you do not know how to speak Italian. Filipino here sometimes are lonely. Sometimes they have no job. Many Filipino are jobless. So they go to McDonald's and ask other Filipino to help them [...] Sometime I go there [the McDonald's], when I have time and am not working, or I stay at home and listen to music. When I listen to music, all my problems disappear. Or I go to the call centre in the metro and phone my husband, or my daughter in Manila, or my daughter in the province. (17 Feb. 2005)

As Maria Anacleta's words indicate, caring for older people as live-in help is a particularly difficult job in many ways. Because of the long, irregular hours, it leaves little free time and no space of one's own to socialize with friends, exacerbating the loneliness many women already feel (Paltrinieri, 2001; see also the Schuster Youth, in chapter 5). Maria Anacleta felt that taking care of an elderly person did not particularly encourage her to learn Italian, making other, different employment harder to attain (see also Paltrinieri, 2001). Because of the age of the employer, moreover, it is perforce a temporary and uncertain position. As Maria Anacleta described, when the elderly person in care passes away, the caregiver finds herself suddenly without a home and without an income and has a very limited time in which to find a new job and a new place to live. In Milan, where there is a chronic shortage of afford-able housing, these situations are especially difficult to resolve.

Working as a live-in caregiver means that the women have to leave their own children behind, becoming "transnational mothers" (Zontini, 2004) for long periods of time. Phone calls and letters are an important way to stay in touch, especially since visits to the Philippines are often hard to arrange. As Maria Anacleta recounted, apart from the cost of travel, many women find it hard to return home for visits because they need to have residence permits in order to be readmitted. The Italian immigration system has been criticized by both the media and activist associations as being one of the worst organized in Europe, making papers hard to obtain and encouraging illegal immigration (Merrill, 2011; Murer, 2003; Leogrande and Naletto, 2002). The difficulty in acquiring papers can be seen as one more way in which Italian society encourages a non-organized, low-paid, and flexible pool of domestic workers. Because the granting of work permits depends in large part on the employer, it makes migrant women even more dependent on the families where they work and with whom they often live.

Maria Anacleta's itineraries in the city reflect some of these issues and difficulties, as well as some of the resources available in Milan. The tours we walked together and the photographs we took, more-over, were a way to forge and strengthen many connections at once: to Milan, to the Philippines, and to other Filipino-Italians in the city. Dur-ing our tours, Maria Anacleta showed me some of the sites she uses in her everyday life, such as one of the Filipino churches, an international calling centre, the McDonald's restaurants, the newsstand where Fili-pino newspapers are sold, a bank catering to the Filipino community, and some of the central piazzas where she spends some time when she is not working. Maria Anacleta guided me to some of the central attrac-tions of Milan, too.

Every time Maria Anacleta guided me in the city, she asked me to take pictures of her in specific places, such as in the middle of a street exhibi-tion in one of the most popular promenades of the centre of Milan, in front of the Duomo Cathedral, and in front of the Sforza Castle and its newly renovated fountain. In fact, most of the visual itinerary of our walks consists of frontal, full-view images of an elegant and smiling Maria Anacleta, confident and at ease in front of monuments, buildings, and piazzas.

Maria Anacleta sent many of these pictures home to her family to show them the city she lives in, where she calls them from, and who some of her friends are. Pictures are one of the many ways to maintain "transnational families" (Zontini, 2004: 1117; see also Parreñas, 2001; and Wolbert, 2001). Zontini describes this as "kin work" (ibid.; see also Di Leonardo, 1987: 442), the myriad everyday practices carried out by Filipina migrants that are crucial in nourishing ties between family members who live far apart for long periods of time. In this context, photographs are often ambivalent. Maria Anacleta's pictures, for example, denote a connection to Milan while also being a tangible sign of displacement.

The photographs of Maria Anacleta in front of some of the major attractions of Milan are, literally, a way to place one's self within the urban landscape. As Mohamed Ba explained (see chapter 8), pictures are an important way to claim one's presence in a (more or less) new city:

> The immigrant who arrives in Italy [...] finds her/his friends who the next day go and buy a roll of film, and where do they bring her/him? Here, in the centre of Milan." (26 March 2005)

These very photographs, however, can easily become a burden for a new immigrant, as they can create high expectations from his or her family members in the home country. Pictures in the city's "status places" veil the difficult conditions most immigrants encounter and the social inequalities that characterize life in Milan.

The photographs that Maria Anacleta and I took together during our walk – including pictures of a Filipino church, of the calling centre she uses, and of places in Milan she frequently visits in order to meet co-nationals – are, moreover, the beginning of a map of the Filipino presence in the city. Public spaces like the central piazzas and some parks, affordable restaurants like McDonald's, and some Roman Catholic churches[3] are important spaces for gathering, especially since Milan offers few other sites where Filipino-Italian people can socialize (Cologna, 2003b). These pictures indicate that, for Maria Anacleta, going around the city is a way to meet and connect with other people originally from the Philippines, who, as she explained in the text above, are a tremendous resource for her and other new immigrants (Cologna, 2003b). Filipina-Italian women often provide support for each other ranging from friendship, to small loans, storage of personal belongings, a place to stay, and help with child care. One of the most important forms of assistance is helping other women to attain employment. Indeed, Maria Anacleta found most of her jobs through the help of friends. Her itinerary in the city, then, is an ongoing, everyday practice of activating and maintaining connections crucial to her survival.

Last but not least, the very activity of taking pictures in the city added new dimensions to familiar walking routines. Equipped with a friend/ anthropologist and a camera, Maria Anacleta could transform a walk in the city into an activity associated with leisure and status. Maria Anacleta pointed out repeatedly that "we are just like tourists!" and just like "those people who have money" (8 Nov. 2004). The very practice of taking pictures in the city enabled Maria Anacleta (and me) to play a different role in the city for a day and to inhabit urban space differently – making it easier, for example, to pose in front of police officers on their horses, as in one of the pictures we took. Marta, another one of my guides, a young woman originally from Peru who combines care of an elderly Italian couple with other jobs, similarly told me that walking in the city and visiting particular sites helps her distract her mind from her everyday life, as it is a complete change from a routine of heavy and long work hours.

Marta's and Maria Anacleta's comments caused me to wonder: Is there "a way to be" a tourist, a Milanese, or a migrant woman in Milan? And where do these practices and imaginaries come from? In other words, how are women's identities constructed through their itineraries in the city and through their practices of recognizing, using, reinterpreting, or rejecting certain ways of seeing?

Francesca

Very similarly to my encounter with Maria Anacleta, I met Francesca by chance in one of the central piazzas, as we were both resting on a bench and watching passers-by. A white, Milanese-born woman, Francesca described herself as middle class. Between November 2004 and March 2005, Francesca guided me three times through the city, showing me her home and parts of her daily itineraries. These often follow a particular routine, which brings her through the major promenading routes, to department stores, churches, and art sites. Francesca explained:

> I love art and history [...] so I make a round of all the churches of the centre – I know them all [...] Oh the things I have seen! (5 Nov. 2004)

Francesca shared her love of Milan's historic centre with me by guiding me through what she called "the heart of the city." This included the
central promenading routes, churches from different periods, monuments, and historic buildings. For Francesca urban beauty was mostly
equated with history and art; however, it is interesting that shops were
an essential part of it, too. During our tours Francesca pointed out stores
with particularly elegant window displays (including jewellers, patis-
series, clothes and fashion retailing), talked about "traditional" and/
or historic shops (including a silver accessories boutique and a phar-
macy), and marvelled at stores displaying stylish designer and fashion
artefacts (such as a ceramic mosaic tiles store and several boutiques).
Indeed, the seamless connection between beautiful churches, heritage
buildings, and shiny window displays was one of the most striking
aspects of Francesca's itinerary and reflected a landscape characterized
by style, affluence, and aesthetics.

Francesca said,

> In the centre of town, one forgets poverty, everything is beautiful, shiny ...
> [T]here is the perfume of money [and] a certain type of people [who add
> to the ambiance]. (12 Nov. 2004)

The "certain type of people," because of their tastes (see Bourdieu, 1984)
associate themselves with certain consumption spaces more than others: stylish, high-quality shops in contrast to more affordable ones:[4]

> If you appreciate certain things you automatically discard others. If you
> [...] love what is beautiful, and high quality, you do not go to Upim [an
> affordable, popular department store]. Because you would not be able to
> find what you like [...] In the Upim there is a certain type of merchandise
> for a certain type of people, who are not me [...] For me also the people
> who go to certain stores ... [are part of the atmosphere of an area]. (15
> March 2005)

According to Francesca, in the centre of Milan, the products displayed in the shops, the (potential) customers inside the stores, and the
passers-by looking at the windows (just like Francesca and me in our
itineraries) are all active participants in the same aesthetic, style, and
allure. Moreover, in a wider perspective, the landscape of a street or a
piazza – with its churches, historic buildings, and its traffic of bodies
and images – adds to the style of the store, just as the latter participates

in the visual feel of the public spaces that it frames (see also Merlo, 2001). Beautiful shops and displays are then not an additional "layer" to streets and piazzas, but rather help organize the very way different people move through city spaces and relate to other passers-by. Francesca clearly sees herself, at least to a certain extent, as part of this landscape of allure:

> I love what is beautiful [...] The beautiful also gives me the joy of living [...] I feel better [...] [with] the beautiful, the clean, the orderly, *l'ultima novità* [the hottest and newest fashion]. (15 March 2005)

Francesca's comments point out how women's engagement with consumption as both subjects and objects of beauty, as onlookers, audiences, and performers, helps mediate women's access to urban spaces (see, e.g., Del Negro, 2004). As Bondi and Domosh (1998) discuss, shopping areas and store windows both provide women with ideas, clothes, and accessories to participate in the fields of vision of streets and piazzas and create "feminine spaces" where women can legitimately be in the city (see also Blomley, 1996; Domosh, 1996; Glennie and Thrift, 1996). This, of course, has not necessarily been empowering for women. The very way in which huge ubiquitous advertising has literally placed the bodies of women all over town, making them into such a central visual feature in so much of Milanese public space, strengthens the suspicion that women more than men need to appear in order to participate in public space (see Ruggerone, 2004). This appearance is always indexed by class and race (Soley-Beltran, 2004), making some women – like Francesca – fit more easily than others into certain landscapes.

Francesca's attention to the aesthetic qualities of the centre of Milan and its economies of looking and appearing is part of a wider concept of beauty, which, encompassing art, tradition, and culture, could serve to identify "Milanese-ness" itself. Indeed, an interesting slippage seemed to occur in our conversations while walking: since historic buildings and artefacts are usually considered beautiful, beauty emerged as a way to recognize and define things historic, precious, and authentically Milanese. Here slippage underscores the complex construction of authenticity through particularly situated enactments, and it has to be understood in relation to the wider social context in which it takes place (Fife, 2004). Francesca insisted that "all Milanese" should do as she does, using every occasion to promenade in the city, and take different routes to discover the treasures hidden in unlikely corners, dark courtyards, and inaccessible palaces. Francesca was particularly proud of her ability to reach historic and artistic sites behind the closed doors of buildings and churches. For her, to know how to access those sites as well as the capacity to understand their beauty is part of living in the city, to be part of its culture and history, and to know how to use its resources to one's own advantage.

What interests me most here is how the very activity of seeing and walking through the city becomes a practice of engagement with urban locales and a way of negotiating one's identity. For it is the repeated, daily practice of searching, looking at, and "appreciating" the "beauty of style and history" that helps Francesca confirm her legitimate belonging in Milan. Here walking and looking make each other possible. Her daily promenading, seeing and recognizing certain landscapes confirms Francesca as a legitimate viewer/speaker, at the same time that her knowledgeable movement through the city puts her, literally, in the position to see and discover art, history, beauty, and style (see Guano, 2003).

Particularly telling in this context was Francesca's frequent use of the Milanese dialect, a language now only spoken by a few people, during our tours:

"This is an umenone," *she tells me, as we stop in front of a large stone statue of a bare-breasted woman who is busy holding the rest of the building on her shoulders. In our tours, Francesca often interjects words and sentences in Milanese dialect. Milanese seems particularly apt for describing Milanese things – such as the umenone. "Do you know what it means?" she asks, since she knows that I do not speak the dialect and thus likely do not understand what this word stands for. "It means big man," she tells me, irrespective of the fact that it is clearly a woman she is pointing to, "They are typical of Milan, the umenones." (15 Dec. 2004)*

The Milanese dialect, once widely used in the city's households, is today often associated with authenticity and with "real Milanese-ness."[5] The very fact Francesca can speak it positions her as an authoritative speaker of things Milanese. In turn, the act of naming and characterizing something as typically Milanese strengthens the sense of specificity of her act of walking through and recognizing the city: it is not just any tour of Milan, it is a tale of a Milanese in the city. Francesca's itinerary is a combination of narrative and vision through which a particularly positioned viewer/speaker privileges one of many possible landscapes.

Francesca's tour "valorized [...] 'certain relationships between people in particular places' [...] thus striving to generate 'consensuses' on these places as well as the identity and entitlements of those who inhabited them" (Guano, 2003: 358, quoting Lefebvre, 1991).

Francesca remarked that foreign-born newcomers are often not in a position to enjoy and appreciate the art and history that Milan has to offer. It is true that many immigrants – as well as many Italians – work several jobs to make ends meet and thus might not have much time to go around the city nor money to pay for tickets to access expositions and museums; nevertheless, Francesca's comments resonate with a commonly held assumption that immigrants do not *really* participate in city life (Merrill, 2006; Murer, 2003; Dines, 2002; Maritano, 2002; Krause, 2001). Many of the people I interviewed, for example, imagined immigrants using *different* shops from "Italians," living in *different* houses, pursuing *different* activities, and ignoring "things Italian" including art, fashion, and culture. A young Senegalese man told me that one of the hardest things about living in Milan was precisely the ignorance of many people regarding immigrants and their countries of origin, as well as regarding cultures and religions different from their own.

Maria Anacleta's photographs in front of churches and monuments, as well as my encounters and conversations with other residents who were originally from other countries, contradicted local ideas that immigrants are not interested in art or "beauty" or are not knowledgeable enough to "see" it. Maria Anacleta told me she liked Italian statues because they reminded her of learning history in school. Marta wanted to see the Scala theatre and took time off work to take me to visit Leonardo da Vinci's *Last Supper*. I did not realize the extent to which I myself took for granted the hegemonic association of Milanese language, history, and identity until Mohamed Ba (see chapter 8) surprised me in recounting the history of Milan in impressive detail, and in citing songs and phrases in Milanese dialect to interpret his own experience as an immigrant to the city.

Connections and Disconnections

One of the aspects that struck me during these itineraries is how Francesca's and Maria Anacleta's seeing practices help delineate social categories and difference in urban spaces by the very way in which they place bodies in landscapes. The connection between embodiment and vision is especially significant here because these itineraries interweave discourses and practices related to fashion, aesthetics, gender, and ethnicity. Visibility plays a role in the way these forces, ideas, and structures come to life in Milanese public spaces. For one, Maria Anacleta's and Francesca's embodied gazes, appearing acts, and visual journeys through Milan reflect, challenge, and reinscribe some of the contours of gender, race, and class that affect both city spaces and the specific ways in which my interlocutors relate to urban locales. For another, they invite us to think that different ways of seeing the city by differently positioned people intersect, reinforce, or disrupt one another, thus taking part in complex visual cultures in which both differences and connections are manufactured and commented upon.

Gender roles and dilemmas in Filipino and Italian society, global and local inequalities, and the racialization of care – intensified by neoliberal privatization – shape the immigration of Filipina women, the issues they face, and the resources and imaginations they seek to harness. In turn, these affect how Filipina women might take part in urban landscapes and how they might be seen within them. Granata, Novak, and Polizzi (2003), for example, point out the particular mix of visibility and invisibility of Filipina-Italian women in Milan. Because many of these women reside with their (middle- and upper-class) employers, they often use and live in the centre of the city *and* they have no place of their own there. This makes them both invisible as legitimate residents and highly visible as migrant workers. To borrow Gordon's words, here the intersections of gender, race, and class determine "the shape [...] [of a particular] absence" in the urban terrain (1997: 6).

Discourses and practices centred on "beauty," style, and aesthetics are another way in which particular gender identities are constructed in and through city spaces. If fashion – as a consumption and leisure activity, as an embodied practice, and as a visual culture – participates in mediating women's access to public space, these practices can strengthen the conceptual division between masculine and feminine subjects. As Joanne Entwistle writes, women's roles as "frivolous" and "foolish" consumers tempted by "the vanities of dress" (2000: 54 and 22) contrast with men's skilful presence in urban spaces as public and political personae. The hegemonic association between gender, adornment, and appearance, which disadvantages women from full political citizenship in public space (Ruggerone, 2004), also obscures that shopping is an important part of women's reproductive labour[6] (Glennie and Thrift, 1996), as well as a source of sociability between women.

Embodied fashion in urban locales constitutes gender in intimate and intricate links with class and race. For one, it is generally easier for white, working-class women to use fashion to pass as middle class than it is for visible minority women. Black women, for example, are often assumed to be involved in low-paying jobs, irrespective of their dress, profession, and social class (Cole and Booth, 2007; Breveglieri, Cologna, and Silva, 1999: 53 and 68) – unless their body shape can clearly identify them as models employed in the fashion industry. For another, style and fashion erases *and* reinforces class hierarchies in very interesting ways. As one of my guides recounted (a young, white, Italian-born, working-class woman), almost anybody can dress well relatively cheaply *and yet* class is still of importance. Although it is not possible to tell directly class from dress, those who can afford expensive clothing and accessories can choose to use them as status symbols. Most importantly, as Francesca suggested, the very *combination* of one's ways of dressing, where and when one promenades, and a practised sense of entitlement to certain spaces reflect and are shaped by one's social position.

The shifting, daily interplays between social categories, and the ways these inform and are shaped by women's relations to urban spaces, inscribe both differences and connections between the itineraries I presented above. The immigration of female domestic workers participates in the very constitution and negotiation of ideals of femininity and gender roles of middle- and upper-class women, which include being successful workers/professionals, charming and affectionate companions, effective housekeepers, watchful mothers, and caring daughters. As Bridget Anderson (1999) argues, it is the work of migrant women that enables middle- and upper-class Italian women to juggle these unsustainable situations and contradictory identities. Italian women, in fact, "buy out" time through the employment of a domestic worker, for "maintaining themselves as 'proper wives' and 'proper mothers'" (1999: 119) without confronting male family members about the division of labour in the house (see also Parreñas, 2001).

Andrea Muehlebach (2012) further argues that the neoliberal privatization of care in Lombardy and its reliance on volunteerism has helped create a deep chasm between moral citizens and domestic (especially immigrant) caregivers who are "merely workers." The latter have emerged as figures not only excluded from, but also in opposition to contemporary reformulations of citizenship. As part of the neoliberal shift in Italy starting in the 1980s, the Lombardy Region decreased support for care through funding cuts, hiring freezes, and privatization.

At the same time, volunteers who offer assistance to those in need (the elderly, the sick, the disabled, the marginal) have been praised and celebrated as a cornerstone of society, and they have become a crucial vehicle for service provision. As Muehlebach argues, this "has allowed for the state to conflate voluntary labor with good citizenship, and unwaged work with gifting" (2012: 6).

Because immigrant caregivers are seen as self-interested workers, motivated only by material needs, they represent the antithesis of the paradigmatic volunteer, who is seen as being "animated not by homo oeconomicus but by what one might call homo relationalis, [...] not by a rational entrepreneurial subject but by a compassionate one" (Muehlebach, 2012: 6). In this way, morality and generosity are constructed as an intimate part of the neoliberal situation, as they nourish its imaginative and material processes, while paradoxically seeming to work against them.

One of the consequences of this process is that immigrant women caregivers like Maria Anacleta, although they perform intimate and necessary care, are excluded from this new model of citizenship based on morality, affect, and relationality. As Muehlebach explains, the "systemic association of immigrant women with material rather than relational forms of production allows for the exclusion of immigrants from the realm of ethical citizenship, that is to say from the sphere of practices that Italians associate with the making of true human relations" (2012: 203).

Muehelebach's work is illuminating here in a number of ways. Tracing wider processes of differentiations at work in Italy today helps us understand how Milanese women from two different categories (the caregiver and the citizen; Francesca is coincidentally also an active volunteer) can be both intimately connected yet stand deeply apart. The visual practices of my two interlocutors that I discuss above embody these disjunctures. The very invisibility of domestic workers is here an interesting case in point. As Muehlebach describes, women like Maria Anacleta are often invisible workers because relegated to the (usually unregulated, precarious, and overworked) private realm, yet they are at the same time "jarring figures" (2012: 219), both in public discourse and in the urban landscape, as they are very often and obviously the ones who accompany children and the elderly[7] through the streets, piazzas, and parks. What is very interesting, moreover, is that their being labelled as self-interested, merely "material" rather than "relational" workers emerges as a cultural dimension, too (212ff). Women like Maria Anacleta do the work of mothers and daughters, yet according to common discourse, they can only approximate these positions because they are "not Italian enough" to deeply relate to the Italian people they care for (213, 220).

Italian-ness emerges here as something both essential and achieved: essential because described as being an innate attunement to Italian traditions, dispositions, and values, and achieved because shaped, attained, and constructed by particular (here, "moral" and "relational") practices. Although this concept of ethnicity and culture echoes more general redefinitions (Comaroff and Comaroff, 2009), it is especially relevant in this case, because it helps us understand the positions of my two guides. Both Francesca and Maria Anacleta walk through and see the city in a way that is strikingly akin to that of a tourist. If neither of them becomes a tourist, however, it is for very different – and, indeed, opposite – reasons. Francesca does not call herself a tourist (although she shows me historic and artistic landmarks of the city) because her very itinerary and its ways of relating to the landscape construct her as quintessentially Milanese. Here walking becomes a practice of achieving an identity as well as expressing its innate characteristics. Maria Anacleta does not really become a tourist either (although she expresses the desire to be one), not only because of her economic difficulties, but also because the very label "Filipina" is associated in this city with caregiver and worker – both constructed through her work and at the same time seen as an essential part of her identity. It is precisely with and against the grain of these categories that walking in the city becomes an embodied, performative way of expressing, negotiating, and reflecting on identities and belonging.

Another interesting connection between women like Maria Anacleta and Francesca is the way in which discourses and practices of aesthetics, tradition, Milanese-ness, and urban renewal affect different women's choices and chances concerning housing. As detailed in chapter 6, as part of her tour, Francesca guided me through the Isola neighbourhood, which, as she described it, was currently being "cleaned up." Francesca, who was planning to move to the neighbourhood, told me that while the restructuring meant that many of the old tenants – mostly immigrants – would be evicted, she was relieved to know that they would find "a new home in one of the popular housing units." In spite of Francesca's best intentions, however, this is doubtful: as I discussed earlier, housing is one of the main problems for people immigrating to Milan, with many immigrants living in cramped quarters or even in abandoned buildings in the dismissed areas.

For Francesca, a certain way of seeing and being in the city legitimizes her move to this neighbourhood and contributes, albeit not intentionally, to making the housing issues of "others" invisible and/or irrelevant. Francesca, in fact, expressed a feeling of entitlement to the area, which she described as one of the oldest and most "authentically Milanese" neighbourhoods in the city, due to her love and appreciation of its historic and artistic heritage, cultivated through her frequent walks.[8] Here I want to stress thay my goal is not to in any way criticize Francesca, but rather to point to structural conditions. Francesca is a generous, kind, and open-minded woman, who has positive feelings towards migrants. I do not intend to say that Francesca's move displaced prior inhabitants; however, I do want to point out the connection between Milanese-ness, a privileged aesthetic, redevelopment, and gentrification. Here, too, attending to daily, embodied practices of vision can help us recognize that different women's locations and journeys through the city and its public spaces are not just parallel but interlinked and connected in important ways.

The juxtaposition of the two itineraries underlines that there are not only many different ways of seeing and being in the city, but also that these interact in complex ways with local and contested imaginaries of who participates in the city and how. Maria Anacleta's and Francesca's circumstances and life stories shape the ways in which they see Milan, walk and talk through it, and create links to urban locales. At the same time, it is important to remember that Maria Anacleta's and Francesca's ways of seeing and being in the city are not mutually exclusive, although they might be imagined to be so by many people living in Milan. Indeed, the very assumption that they be so is an important aspect of the negotiation of claims to the city on the part of those who consider themselves Milanese (in an exclusionary way). As the Schuster Youth and Don Felice pointed out, imagining that "non-Italians" in Milan use public space essentially differently from the "Milanese" can be used to legitimize stereotypical perceptions of migrant women and men. It can contribute to the "struggle to exclude the migrants, who are [seen as] 'taking over' the buildings, the neighbourhood and the city" (Maritano, 2004: 69).

Just like stories about who participates in the *struscio* and who does not, the itineraries of Maria Anacleta and Francesca suggest that tales of immigration and of aesthetics are both integral parts of who can participate and how in public space. This is because these discourses shape the contexts and spaces in which the two women could meet and interact and the ways in which their itineraries could cross. Paradoxically, in fact, Maria Anacleta and Francesca are not likely to meet in the centre even if they use very similar streets and piazzas. Indeed, the discrimination and avoidance practised by many Italian-born residents in Milan contradicts the ideal of public space as a place where "one always risks encountering those who are different" (Caldeira, 2000: 301), that is, a "public space founded on uncertainty and openness" (303) and on "difference without exclusion" (Young, 1990, quoted in Caldeira, 2000: 301).[9]

Although this seems like a pessimistic note on which to end this chapter, we should not underestimate the very possibilities of alternative imaginings, visions, and spatial strategies women engage in through their daily lives. The very differences between Maria Anacleta's and Francesca's pictures can be an interesting example in this respect. Francesca was never depicted in the photographs we took during our walk. This absence from the photographs can be seen as a sign of power: it underlines that she occupies the legitimate viewing position. Yet, this absence could be interpreted as denoting a sense of loss of her position in the midst of changes to the city. Conversely, while Maria Anacleta's pictures in front of monuments place her somewhere between a visitor and a Milanese, her strong visual presence in the city reflects the ways in which immigrant women are shaping, claiming, and constructing urban spaces and demanding justice and recognition.

Chapter Eight

Places and Stages: Vision and Performance in Public Space[1]

Theatre is an itinerary in hope.

– Mohamed Ba

This chapter follows Mohamed Ba, a community educator and theatre writer, as he guides me through the centre of Milan. I was introduced to Mohamed Ba by the staff of the street newspaper *Terre di Mezzo*, who I had interviewed some time earlier. Mohamed Ba and I met on a bitter cold Saturday morning (26 March 2005) in the Duomo Piazza: like with Francesca and Maria Anacleta, I had asked him to show me "his" city, and the following itinerary, journey, and images are the results of this walk we did together.

When Mohamed Ba arrived in Italy several years ago from Senegal, his work was with *Terre di Mezzo*, both as a vendor and as a liaison person between the vendors and the editorial board. His tour and perspective on the city centre reflect this experience. For one, he describes how a recent immigrant who works as a street seller might navigate his way through Milan's core. For another, Mohamed Ba talks about the streets themselves as an avenue for knowledge, sociability, and ultimately hope, as the vendors and passers-by (typically immigrants and Italian-born residents) use public spaces to meet and learn about each other.

According to Cologna, Breveglieri, Granata, and Novak (1999), selling in city streets, piazzas, and markets has been one of the most important occupations for Senegalese migrants, who have been arriving to Italy in significant numbers since the beginning of the 1980s. As Mohamed Ba describes below, the co-nationals who help a newcomer settling in, often offer him (the migrants are mostly young men; see Cologna et al., 1999) a collection of wares to sell as a first step to establishing his business; in addition to selling, Senegalese-Italian residents of Milan and its metropolitan area are increasingly being employed by small and medium-sized industrial businesses in the region (44). Cologna and colleagues explain that, throughout the 1990s, the work opportunities in this field, in fact, made Lombardy into an important area of settlement for many people from Senegal who had initially moved to other regions of Italy. A young Senegalese-Italian man I met in Milan, for example, described how he alternated working in a manufacturing firm with some of his cousins and selling books in the streets when there was not enough work for him at the factory.

Talking about the difficult process of settling and working in Milan, Mohamed Ba's itinerary through the centre that I present in this chapter encompasses much more than a description of a newcomer's experience of the city. By adopting at the same time the perspectives of a long-time Milanese resident and a new immigrant in precarious conditions, my guide problematizes easy categories of "insiders" and "outsiders" to the city. Nor are these two positions the only ones that inform his tale. During the tour, Mohamed Ba becomes a time traveller in the past and the future; an elder and a young man; a scriptwriter, narrator, and performer/actor; a teacher and a questioning student. One of the aspects of Mohamed Ba's itinerary that interests me most, moreover, is its starkly theatrical character. He crafted a wonderfully elaborate and artful monologue that included proverbs, poetic verses, choruses, and even a song, and that moved effortlessly from one piazza and/or street to another. During the tour, I felt like a spectator following a representation of the city through changing and interlinked scenes. Similarly to Richard Freeman's (2001) description of Buenos Aires, the city itself emerged as a "mise-èn-scene," and its very locations as stages for (a) performance(s).

In this chapter, I argue that the performative sense of Mohamed Ba's tour is particularly important in constructing alternative notions of "belonging" to Milan and in enacting public space as a creative site for social transformation. Critical acts of imagination and vision enabled by performance are crucial tools in his project. By engaging the streets and piazzas of Milan as stages, Mohamed Ba's itinerary interweaves urban landscapes, real and imagined life stories, and a multitude of speaking/viewing/walking positions. Because his narration and itinerary construct such a complex location for a cultural commentator, they challenge assumptions that there is only one category of Milanese – or of Senegalese, or of migrant – and that only people born in a city can truly know and understand it. Imagination here emerges as a "social practice" that is "central to [...] agency" (Appadurai, 1996: 31), a key element of social life as it can direct one's attention, shape individual and collective projects, and spark a questioning of social reality.

In attending to Mohamed Ba's tour as a theatrical act, I find Bauman and Briggs' (1990) discussion of performance useful. They point out that performances have to be seen as deeply tied to contexts, such as discourses, situations, relationships, and/or other performances that follow or precede them. At the same time, performances are particularly apt at transcending those very contexts and thus creating "memorable text[s]" (73), which "can be lifted out of [their] [...] interactional setting[s]" (ibid.) and can then again play a role in other situations. Indeed, the performative form of Mohamed Ba's tour ensures that his audience experiences and remembers it as a "memorable" and authoritative text on the city and its history. Not simply a walk through the city, it is a moment of teaching – a commentary that aims at reframing landscapes and experiences to foster understanding. According to Mohamed Ba's words below, the chance to do just that, to talk about the city and act as its guide, is a powerful way to show that he belongs to the city and that the city also belongs to him. To say it differently, "performance puts the act of speaking on display" (73). As such, it emphasizes that, in the context of Milan's emergent multiculturalism, the very act of talking about the city by a speaker positioned from between and within cultures and places (see Tsing, 1993) is a significant political and pedagogical praxis.

Because the performative form of Mohamed Ba's itinerary was important, I have chosen to represent it as a theatre play (I found Madison's 1999 work inspiring in this regard). To do so, I have presented long passages from Mohamed Ba's narration and linked the excerpts to the urban "stages" he used. I quote Mohamed Ba at length to show the movement between different speaking positions. Most importantly, however, I am interested in creating a space for his intervention as a critical, counter-commentary on Milanese public space, an important "back talk" in relation to contemporary discourses on immigration and multiculturalism in the city. Following Kathleen Stewart, I am attempting to use "the possibilities of narrative itself to fashion a gap in the order of things" (1996: 3).

I represent the setting both through pictures and words (the stage directions) because the location in and journey through actual city spaces was an important referent for Mohamed Ba's tale. Note that the images and the written descriptions do not always match. The pictures make visible to the reader the places (albeit not all of them) where Mohamed Ba told me about "his" Milan, and my descriptions adapt those settings to a possible theatre stage. They also introduce the following elements that were not part of our walk: the choruses (in most scenes), two men with masks (in scene III), five characters in the shadows (in scene III), and the typewriter, chair, table, canvas, and board (in scenes I to III), with the associated actions of creating and sharing texts and maps.[2]

The discrepancies between the pictures and the stage directions thus reflect the particular status of this text, which is neither (or both) a verbatim transcription nor (and) a fictional writing. All of the lines of Mohamed Ba (and my few brief comments within them) are translated quotes from the transcript. The itinerary presented in the play largely corresponds to the route we took through the city (I indicate in the text when the two differ from each other). The street vendor, the woman with an accordion, the passers-by with shopping bags, and the sound of the church bells were all present in our tour through Milan. However, I have added fictional moments and use a theatre script genre because I believe that this form evokes, better than a regular academic discussion, the movements through the city, the sights and sounds, and the creation of affect[3] that are so central in Mohamed Ba's commentary. As Nigel Thrift (2003) points out, these are aspects that are lost in usual scholarly discussions, although they are constituent parts of all our interaction with the world and with others.

In the text below, I have also inscribed myself as a participant in the play. This reflects my double role, as both a listening/following spectator to whom the tour is directed and as a particularly positioned commentator who writes the itinerary into a text, thus interpreting and representing it for other audiences. My first goal in becoming a character in the writing below is to point out how Mohamed Ba's intervention created a significant space for an audience. His very description of theatre in Senegal underlines that the spectator is always an active and fundamental part of the play:

> The theatre is not that spectacle where there are, on one side, actors and, on the other side, the public. No, because [...] the public who listens or follows the story becomes automatically the protagonist of the story it listens to.

Although I say very little throughout Mohamed Ba's narration, and could hardly be called a protagonist, I still had a specific task to carry out: to witness the story and be transformed by it. Performing can put not only "the act of speaking on display" (Bauman and Briggs, 1990: 73), but also the act of listening. As Jill Dolan suggests, theatre can intensify the act of listening to another person, with the aim of "model[ling] a hopeful method for living near others with respect and affection" (2005: 88).

It is important here to note that this itinerary is a version of a walking tour that Mohamed Ba prepared and performed for a group of *Terre di Mezzo* readers a couple of years before our undertaking. It was part of a small series of guided tours, which aimed to show Milan from a range of different perspectives to interested city residents (mostly white, Italian-born residents). As such, this project was meant to be transformative while directly involving its listeners as the protagonists of this change.

Indeed, during our tour, I was changed. Surprised that Mohamed Ba knew so much about the history of Milan, and that he knew expressions in Milanese dialect that I had only heard from older, Milanese-born residents, I had to confront my own stereotypes about who knows what about the city. But that was not all. The stories he told me, and their rhythm, sounds, sights, and sense of direction and movement moved me, affectively, to imagine what public space could be like. Performance can be a unique tool for social critique and transformation, in that it allows audiences to experience "what utopia could feel like" (Dyer, 1992, quoted in Dolan, 2005: 39). What I learned from Mohamed Ba, in turn, allowed me to look differently at other moments of urban life, such as the events I describe in the latter part of this chapter. This is the second way in which I act as a "participant" audience. As Bauman and Briggs explain, "even when audience members say or do practically nothing at the time of the performance, their role becomes active when they serve as speakers in subsequent entextualizations of the topic at hand" (1990: 69).

Writing myself as a character in the play is, then, a way for me to remind the reader of my own activities of "entextualization" (presenting Mohamed Ba's walk as a detachable text) and especially of "recontextualization" (using this text as a new frame of reference for another series of events/performances). In the second part of this chapter, I use Mohamed Ba's text as a frame of reference for looking at a series of rallies that happened in the centre of Milan. Although a guided tour of the city seems at first sight to be not comparable to "city wars," as these rallies were dubbed in the press, Mohamed Ba's carefully crafted tale of fluid identities reminds us that memory, vision, and performance have powerful implications when they come to life in the streets. Not only does his narration remind us that practices of seeing, representing, and spectatorship are important loci for the creation of political and social identities. By imaginatively placing sociality, cultural creativity, and hope as part of public space, Mohamed Ba's commentary can become one of those memories we can "seize hold of [...] as it flashes up at a moment of danger" (Benjamin, 1969: 257).

A TOUR OF THE CITY

Scene I. The Duomo Cathedral

Figure 26. The Duomo Piazza (*top*; photograph by Aurelio Bonadonna, April 2014) and the Duomo Cathedral (*bottom*; March 2005).

(The Duomo is in the background. It is mostly covered by a white cloth. People and pigeons are walking on the square. Mohamed Ba/the tour guide and Cristina/ the anthropologist are standing in front of the Duomo. To the side of the piazza there is a chair and a table with a typewriter on it. Behind these hangs a white canvas. The typewriter, chair, table, and canvas remain there until the end of scene III.)

MOHAMED/THE TOUR GUIDE *(Standing facing the Duomo, and occasionally pointing to it)*:
The immigrants today see the Duomo as being so majestic ... But in reality it is nothing other than [...] the realization of a very long and tiresome itinerary which has involved the city of Milan for centuries and centuries [...]

So anyone who arrives often looks for the centre of the city in order to orient him(her)self, because often, by searching for the centre of the city, (s)he will be able to move: (s)he enters and exits. And the one who follows well the history of Milan will understand that it [the city] has a belt: the Navigli canals.

So the immigrant[4] [from sub-Saharan Africa] who arrives in Italy is often welcomed by his (her) co-nationals [...] [(S)he] already has a point of reference, a point of approach, that will be The House. And when (s)he comes to that House (s)he will often be struck by its aspect, because one always expects to see houses perhaps very beautiful, with a room for each member, and so on, and one finds oneself in a one-room apartment with fifteen–twenty people, forced to sleep in turns.

That impact with the reality is often embarrassing, but one does not have the right to look back. So already who leaves leaves and the adventure starts from there. But *(voices from backstage join Mohamed's voice, creating a chorus)*[5] the adventure passes also through knowledge: until we know, we won't be able to respect or [...] to appreciate *(end chorus)* because usually one recognizes oneself in the positive values of all cultures.

So our poor immigrant friend from Senegal or from [another] part of Africa, finds him(her)self in Italy, (s)he finds him(her)self friends who the next day go and buy a roll of film, and where do they bring him (her)?

Here, in the centre of Milan. (S)he sees this majestic [Duomo] – that seems from far away a porcupine –

(Cristina/the anthropologist laughs.)

it is a symbol of a city.
So, the pigeons that fly for us[6] have become banal, familiar. But for who arrives: (s)he has her mouth watering, why?

Because they are eaten in our side [of the world] and (s)he already imag-
ines a barbeque with many pigeons, and so on.

(Cristina/the anthropologist laughs.)

And (s)he is told: no, no, no, one does not touch them, one does not eat
them. So resignation sets in. So (s)he tries to pose him(her)self questions:
But this city was always like this?
How come it became like this?
This piazza, what does it represent to the citizens?
But unfortunately his co-nationals are not able to explain. The
explanation that (s)he is told is that here is the meeting point: when the
Milan [soccer team] wins the tournament one meets here, at New Year's
Eve one meets here, perhaps even in the past one assembled here. So
another[7] itinerary begins, through the belt of Milan that is the belt of the
Navigli [canals].

The majority of countries of sub-Saharan Africa suffer from drought; it is a
zone where it rarely rains. [In Senegal] colonization introduced an industrial
monoculture of peanuts [...] The earth has suffered. At the end there is an
advancing of the desert [...] and the farmers have been forced to go to the urban
centres [...] And then [to] Europe, following a dream, legitimate even, to have a
better life, to have the flexibility to get up in the morning and to start to dream.
But a Europe particularly rooted in our habits and customs because through
colonization we acquired a double cultural identity [...]

And so one closes one's eyes and leaves. One throws oneself, one
throws oneself in this city: big, majestic. And when one finds oneself
here to snap 78 photographs that one sends home the next day, with
pigeons, this gives a bit of serenity to the family that was anxious for
his (her) departure [...]

But this is only the first step. Slowly (s)he follows the course of the water
and (s)he asks him(her)self: but water [here] is so sterile, nothing moves, it
seems almost in winter sleep. Why? Why? Because fortunately water is not
something lacking on this side [of the world], but on the other side, where
water is alive, everyone ... animals, caravaners, everyone meets around the
water [...]

Following the course of the water, (s)he realizes that it becomes a belt,
a belt that suddenly ends behind here[8] [...] and (s)he tells him(her)self,
but how come (s)he ended in a street which on the other side is called Via
Laghetto [Little Lake Street].

So (s)he poses him(her)self the question: [...] Via Laghetto? It is strange,
because here there isn't any water!

But yes, there was water: water underneath.

But what was that water used for?

So (s)he continues his (her) tour and looks and cannot see another way than the Duomo[9] [...]

The Duomo is a factory that never ends. If one in Milan is taking a really long time to do something, often one tells him: you are slow like the works of the Duomo. Because from the remote times, when the work began, the construction is never ending [...]

(Cristina/the anthropologist walks away from the Duomo square, as the lights go out from the Duomo square. She sits in front of the typewriter, and starts writing this text. The text is projected onto the white canvas hanging behind her.)

LANDSCAPES, JOURNEYS, AND VOICES

Landscapes

Mohamed's descriptions and reflections remind me that we can think of landscapes as the interweaving of the visible and the invisible, the "given-to-be-seen" (Taylor, 1997: 122) and the hidden, the details that are present and what is absent. In his words, and in the journeys of an unknown (or perhaps even too well known) friend, the web of the Navigli canals shines through the city floor behind the cathedral, while the Sforza Castle (see below) becomes opaque, forever closed, and incomprehensible. Listening to him while looking at the Duomo, I notice that the cover, more than concealing the face of the cathedral, makes starkly visible its history and reputation as a never ending factory and construction zone. Animals come to the foreground, from the pigeons imagined on a barbeque (a theme developed also by Calvino's fiction[10]) to the half-woolled pig that Mohamed talks about below.

What comes to be on which side of the divide between visible and invisible, present and absent, depends in important ways on the position of the person watching, moving, and being within a landscape. Mohamed points out that the perspective of a viewer is itself never simple. On the one hand, it encompasses multiple, sometimes even contradictory visions (see, e.g., Rose, 1993). It includes not only what we see, but also what we are supposed to see. It includes "the shape" of the "absence" (Gordon, 1997: 6) of what could be there, and as such lives in the space of dreams, wishes, or hauntings. The "house perhaps very beautiful," the pictures of the Duomo, the pigeons, the water: they all lend themselves to being seen at least twice, from different perspectives. They look different for different social actors, because they occupy different places in people's experiences, expectations, and daily lives. Mohamed suggests that it is by "looking twice" and adopting different points of view that we can, for a moment, be different subjects within the landscape.

On the other hand, it is by moving through the city and following its routes that a particular vision becomes possible. For the recent immigrant, the waterscape of the Navigli, now mostly covered by asphalt, emerges through a journey of discovery of the new city she or he finds herself or himself in. In this manner, a way of looking connects several itineraries and stories. The walk through the city started, in a sense, from his or her departure from Senegal, and is thus deeply connected to colonization, to the history of sub-Saharan Africa and its waterways. In turn, the way in which "our friend" walks through the city and what she or he sees uncovers other journeys and perspectives of city spaces. It is perhaps in this sense that the Duomo and its piazza are "the realization of a very long and tiresome itinerary" by a multitude of people "which has involved the city of Milan for centuries and centuries."

(End of scene I. Lights off.)

Scene II. The Name of the City

(Mohamed/the tour guide and Cristina/the anthropologist are back in the Duomo Piazza. This time, the Duomo is in the corner of the stage, and next to it there is a crowded street, leading to a low brick and stone building with arches. On the other side of the stage, there is still the typewriter, chair, table, and canvas.)

MOHAMED / THE TOUR GUIDE (*Still standing in the Duomo square, but facing away from the cathedral*):
But the name of the city, where was it born [where does it come from]? So there are many myths and many legends, one of which narrates that there was a tribal chief called Bellovoso, who crossed the Alps and came to Milan, and so everyone asked him what the name of the city was. Not knowing what to answer he asked his councillors [...]

(Mohamed starts to walk away from the square, through a crowded street; Cristina/the anthropologist follows him. People on the street are walking in the opposite direction to them.)

So his councillors went on a retreat for some days – the legend narrates that these days were five. And that is why the number five became important for the city of Milan, because [for example] to liberate this city –

Figure 27. Mercanti Street. (Photograph by Aurelio Bonadonna, April 2014).

CRISTINA/THE ANTHROPOLOGIST: Yes!

MOHAMED/THE TOUR GUIDE: the battle lasted five days, so that's why [there is a] Piazza Five Days, but that is a recent story. So he was told that [...] the city had a name and a symbol, and that [...] the name of the city figured in the symbol, that was a little animal [...]

 So they went around in the city to look for that animal, and it was hard to find, because you can imagine how the city was then, with all these streets always full of people [...]

(Mohamed indicates the street they are walking on with ample gestures.)

And the legend narrates that they took this street that is called Via
Dante which is by Piazza Mercanti [...] so they came under this
palace. And there, where there are the stairs, they found a particular
animal [...] it was [...] the female of the pig, but it was particular
because it had half [of its body covered by] wool – from Latin *media
lanuta*, which later became *Mediolanum* and Milano of our days. And,
in fact, the symbol of that small animal, the pig, can be found right
here.

*(Mohamed stops under the brick and stone building, and points up to one of the
arches, where there is a small carved animal figure.)*

CRISTINA/THE ANTHROPOLOGIST: Ah, that one!
MOHAMED/THE TOUR GUIDE: [...] This is why the history of such a big city
as Milan becomes difficult to understand, because one usually expects to
see a symbol [...] which would be visible to all. But the pig is here, stuck
in this way. So the people ask: but if I have to look for the symbol of
Milan, where do I go to look for it? So the symbol was chosen that would
be visible to all [...] that is the one of the *Madonnina* [the "little Madon-
na," the golden statue of Mary placed on the pinnacle of the Duomo].
*(Mohamed points to the Duomo in the distance, and starts to sing "O mia bela
madunina" in Milanese dialect).*[11]

*(Cristina/the anthropologist returns to her typewriter while Mohamed/the tour guide
continues singing "O mia bela madunina." When the song ends, the lights go out on
the crowded street and the building with arches, and Cristina/the anthropologist starts
again writing her text, which is projected on the white canvas.)*

JOURNEYS

Just like the water of the Navigli, Mohamed's words create a belt, an intricate
net of itineraries through Milan. These are some of the paths that emerge
from his descriptions.

*(Cristina/the anthropologist gets up and starts unrolling the following maps, which
she pins to a board behind her. The maps are made of transparent paper and as she
pins them one over the other the drawn-in itineraries of each map overlap and add to
each other.)*

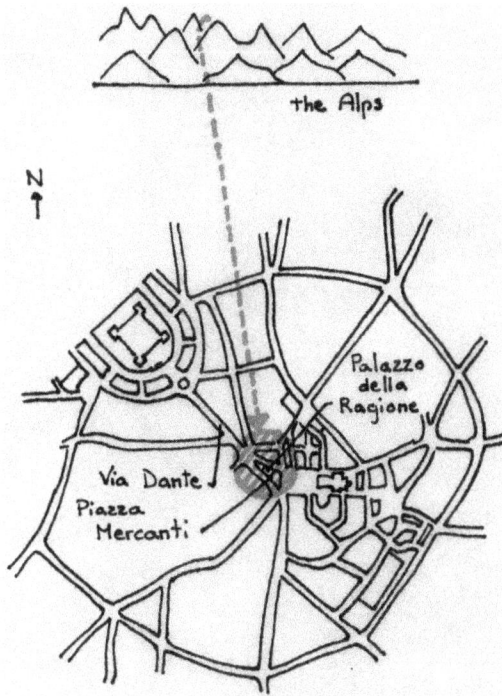

Figure 28. Map of Bellovoso's journey.

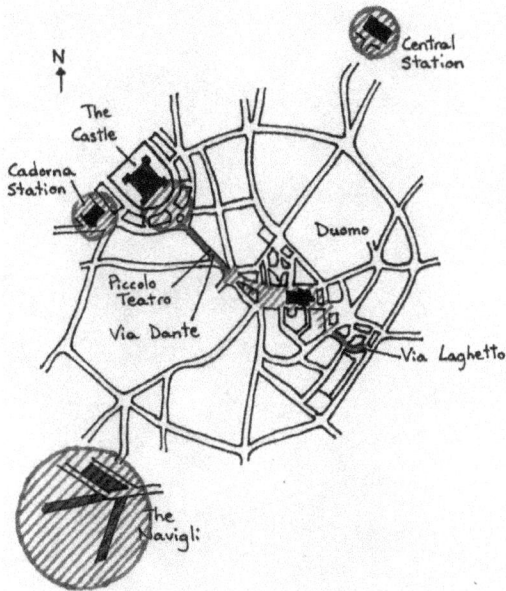

Figure 29. Map of a street vendor's journey.

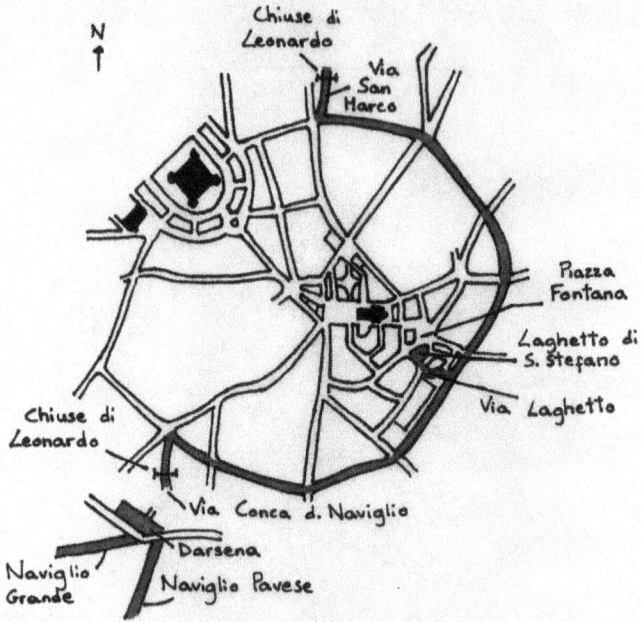

Figure 30. Map of the Navigli canals.

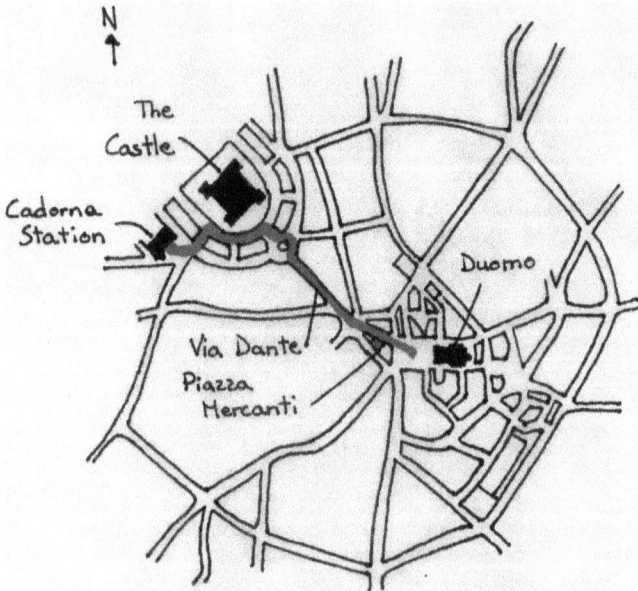

Figure 31. Map of Mohamed/the tour guide and Cristina/the anthropologist's journey.

(Cristina/the anthropologist returns to her typewriter and continues writing.)

These itineraries are not only physical journeys through city streets and piazzas; they span continents, languages, and times. Suddenly it is not so straightforward to tell the history (official or otherwise) and stories of Milan – and we might start to suspect that it never was. To understand it, to tell it, it is necessary to talk about colonization in Africa, as well as about the fascination for Leonardo by European tourists, the legend of Bellovoso, and more. In a way, Mohamed's narration makes space for "ghosts": people, events, and places that are not visible, yet still have effects in present-day life (Gordon, 1997). His comment on the history of Milan being so difficult to understand because it is hidden, "stuck in this way," can be read as an illuminating description of the problem of historical memories stuck between the thick arches of power.

Milan is indeed a city of ghosts, unsolved puzzles, and contradictory memories. Just a few minutes walk away from where we stand is Piazza Fontana, where a bomb killed 16 people in 1969. Still nobody knows who planted the bomb near a busy bank. It is generally believed that it was far-right forces, in a "strategy of tension" aimed at keeping the left from power. But who exactly was involved? Until now the courts have been trying to find an elusive truth. And how did Pinelli die, the anarchist who was being questioned by police about the bombing? Did he really just trip and accidentally fall from a window while in police custody?[12]

One of the interesting things about itineraries is that, if a walking tour through the city can evoke stories, events, and relationships, the opposite is also true: words, tales, and performances conjure up streets and places, and make it possible for us to see, know, and ultimately move through the city. In Mohamed's text, questions – such as: "This city was always like this?" "This piazza, what does it represent to the citizens?" "Why Via Laghetto?" "Why is the city called Milano?" "Where do I look for [the symbol of the city]?" – are the centrepieces of the narration. These questions, however, are also always tentative directions, steps in an itinerary, literally the beginning of streets. Just like the course of the Naviglio tells a story, the story of the migration of a person traces a journey in city spaces.

(End of scene II. Lights out.)

Scene III. Piazza Mercanti/Palazzo della Ragione

Figure 32. Piazza Mercanti. (Photograph by Enrica Sacconi, March 2005).

(Mohamed/the tour guide and Cristina/the anthropologist are standing in a quiet, old courtyard, with what looks like a brick and stone well in the middle. Leaning on one side of the well, there are two young men wearing masks (one mask is black, and the other is white) and playing several musical instruments. To the side of the courtyard, there is still the typewriter, chair, table, and canvas.)

MOHAMED/THE TOUR GUIDE: here was also the place where [...] people came to be judged. In those times bankruptcy was a shame, not only for the artisan but also for his whole family [...] So we find some analogies with the African tradition where honour needs to be defended at all costs [...] If one failed his duty he was not put in prison but brought into a piazza and the elders hit him with words where even the most powerful war tank could not touch him: in his honour, in his dignity [...] But also in Milan when one failed, he was brought into this piazza.

The one who arrives here thinks to have found a well, but in reality it is not a well, because inside there is no water, inside there is a stone [...]

CRISTINA/THE ANTHROPOLOGIST: (*surprised*) A stone?

MOHAMED/THE TOUR GUIDE: inside there is a stone called the stone of the beaten. Why? Like in the African tradition, [in Milan] the one who went bankrupt was brought here, naked, and had to hit the stone three times with his bottom to be shamed in public. And so, you see, (*voices from backstage join Mohamed's voice, creating a chorus*) even if the world seems so old, the beginning of the future emerges always from the past (*end chorus*) [...] One is very sorry if (s)he lives in a city and cannot be the flag carrier of this city when (s)he leaves. Because my Milanese-ness did not detract anything from my African-ness, it has confirmed it, even [...] So we are convinced that the world goes how it does because we have forgotten the weight of culture [...] (*voices from backstage join Mohamed's voice, creating a chorus*) culture is the only thing that is left to a person, even if everything is taken away from him [...][13] (*end chorus*).

But my grandfather told me "when you will happen to go to another country, with other people, if you see that everyone runs after the having, let them go and run after the knowing because sooner or later it will be you who will have to manage what they will have found." And doing the touristic guide today in Milan for me is a payment; my richness and my treasure are to know this city, to appreciate it, in [its] symbolic and its imaginary places.

Because it is a city that has always been contested – in the Middle Ages as well as after; but also in our days, because the foreign communities divide among themselves the Duomo Piazza. There is a corner where only Peruvians meet, on the other side only Africans – when the good weather arrives they go to play drums, bongos, and so on – on the other side Latin Americans, and so on. And the Italians find themselves there, easily consumed [lit.: eaten] by the artistic and cultural expression of those who arrive. Because when I arrive with my djembe, with my drum that I start to play, often the young Italian guy who looks at me gets excited and wants to do what I do. And with desire and passion he will even succeed, but he will forget that he himself has the tarantella, the pizzica [traditional music and dance forms from the regions of Puglia and Basilicata],[14] the "Tammurriata Nera" [an old song from the city of Naples].

So in that context immigration becomes a problem, not a phenomenon, because it will confirm the uprooting of the Italian people from the foundations of its culture, its history, and it is a problem to face. But if the Italian who comes to see me is conscious of having the tarantella and the "Tammurriata Nera," it will be enough to bring it around the table and together we can do interculture. But we can't think about integration by asking one to do exactly what the other does in order to count [that is, to matter in society], otherwise [...] that becomes assimilation.

(Cristina/the anthropologist returns to her typewriter to the side of the setting, as the lights go out on the courtyard. She continues writing her text, while the music continues.)

Voices

What particularly strikes me in Mohamed's play is his use of different voices or perspectives within one tale. I can imagine all of these characters here around me, as I try to distinguish them and to make their acquaintance.

(Cristina/the anthropologist gets up, takes a piece of paper with the text of Mohamed/ the tour guide's narration, and cuts it with scissors into five sections. In the meantime, five figures appear standing in the shadows, on the side of the typewriter. She gives one section of the paper to each of them, as she greets them, one by one.)

CRISTINA / THE ANTHROPOLOGIST: *(addressing the first figure)* You are the narrator?
(addressing the next figure) The voice of history?
(addressing the next figure) Our friend!
(addressing the next figure) Grandfather?
(addressing the next figure, but the audience cannot hear what she says) ...

(She returns to the typewriter, and continues to write her text. The five personages disappear into the darkness.)

Steedly points out how the author of a story always writes himself or herself in it as a "figure in the carpet" (Steedly, 1993: 20). I find this image helpful while I relisten and rewrite Mohamed's tale. Although Mohamed's telling seems at first sight a linear description and a monologue, part of his performative aspect comes from the fact that it includes several characters that are placed in a dialogue with each other. Sometimes this dialogue erupts in the open, creating small vignettes, but often the interaction of these voices is similar to the subtle weaving together of several strands and colours. For this reason, to continue with the analogy, it is often difficult to distinguish exactly one thread and/or pattern from the other. This entanglement is one of the strengths of this performative commentary. By being many and interlinked figures in the carpet of the narration, Mohamed then addresses the ways in which these differently positioned social actors co-inhabit a city like Milan, and how their being together in public spaces creates both tensions and possibilities.

One of the most powerful "doings" of a tale is often the creation of personages that are too slippery for dominant tropes and discourses to anchor themselves on them (Tsing, 1993). Although Mohamed talks about "Senegalese" and "Italians," he carefully presents them as much more complex categories than the ones often imagined in the media and daily discourses. A Senegalese can be a wise grandfather, a recent immigrant, a person with "a double cultural identity." Similarly, an Italian can be a racist, a youth who "forgets" her or his Italian-ness, somebody who changes and learns by engaging in dialogue. Both a "Senegalese" and an "Italian" can be a Milanese: somebody who lives in the city, regardless of his or her nationality and/or colour. Indeed, I find that the strength of his story comes not so much from the fact that it includes many points of view, but in the way in which they shift, mingle, and become one another.

By talking about identity as several dresses, speaking positions, and intersecting itineraries, he points out that "Milanese-ness" does not preclude "African-ness" and vice versa. This also becomes a comment on the very character of urban spaces. The use of the Duomo Piazza by members of different communities is consistent with its history as a meeting place, with the geographical and historical position of Milan as a middle space between trading routes. And again, for "the one who follows well the history of Milan," its thick layer of histories becomes a resource, a "richness" and a "treasure" and a way to become a "flag carrier of a city."

(End of scene III. Lights out.)

Scene IV. The Sforza Castle

(Mohamed/the tour guide and Cristina/the anthropologist are walking on another crowded street. The street is lined with shops. Many of the people walking by carry plastic shopping bags. On one side of the street, a Black man wearing a heavy jacket and a woollen hat is trying to sell books to passers-by. At the end of the street, in the distance, there is the Sforza Castle. The side where the typewriter was placed in the previous scenes is now in the darkness.)

Figure 33. Sforza Castle. (Photograph by Aurelio Bonadonna, April 2014).

MOHAMED/THE TOUR GUIDE:
[...] Our immigrant friend will [...] need a minimum of 8–9 months for him(her) to be able to move around alone. (S)he will ask passers-by, [...] (s) he will get explanations, but not for everything, why? First, time is short, and the second reason is that the one who leaves one's country can have on his (her) shoulders more than forty mouths to feed. Thus, the time for discovery cannot be too long, and everything (s)he will know will be reduced to the necessary minimum [...]

So our friend [...] will be entrusted to somebody else who will act as his (her) tutor, from the same household where (s)he resides. And this person will have the task to help him grow in his (her) work. Everyone who arrives does not speak the language, perhaps does not even have documents and so forth, the only thing that (s)he can do is sell. Sell what? [...] The first day [in the house] (s)he will have to meet everyone, explain how relatives are, and so forth, and then there is a collection. Every member of the household gives him something, perhaps a packet of CDs [...] some t-shirts to sell [...] The next day (s)he will be assigned to a tutor who will bring him (her) along when (s)he goes out selling. Perhaps (s)he will put him (her) a hundred metres from the fixed location where (s)he stays, perhaps in a parking lot, or in front of the stadium, or here in the centre. So for a month, for a month (s)he will not have to pay anything [...] This will give him (her) the time to sell and to put away money, and construct his (her) capital. From the second month, (s)he will have the responsibility to do exactly what the other people in the house do, and (s)he will become a complete member of the household in every respect.

So from the Duomo looking at this road, Via Dante, that brings us directly to the Castle [...] [our immigrant friend] passes by the Piccolo Theatre, but (s)he does not even look at it, because [...] the theatre how it is understood in Africa has nothing to do with the Piccolo Theatre, or La Scala, no, let's forget about it. Because for us the theatre has to give us again the joy of living, the theatre has to be a moment of freedom, of artistic expression, the theatre is not that spectacle where there are, on one side, actors and, on the other side, the public. No, because the public also has to interact, the public who listens or follows the story becomes automatically the protagonist of the story it listens to [...]

And so I remember well the first theatrical performance I did in Italy, with a very good producer of the Teatro Officina [...] so he asked me for the script.

And I said: which script?

But, the script of the performance!

But no, there is no need to write it, it is *my* performance.

Yes, I know it is yours, but I need to know how it will be.

If you want I can narrate it to you.

No, no, you do not need to tell it, you need to write it.

[...]

So I was not ready to do interculture because I had not understood his reality [...] So from there I understood that leaving my country I would have had to borrow a new dress and that it should not at all be tight for me. I just had to realize that I was borrowing it. And this to make better my permanence in Italy. Because we will never be able to communicate if we are speaking different languages [...]

(A woman playing an accordion appears on stage/on the street. She plays slow and repetitive music while Mohamed/the tour guide continues to talk. While playing, she crosses the stage until she exits on the other side.)

This is helpful for me in order to – why not? – learn the positive aspects on this side and perhaps bring them to the other side that I left and see them also grow with a new dress, that they will not have borrowed, it will be always theirs, but it will change a bit their point of view. (*Voices from backstage join Mohamed's voice, creating a chorus.*) Because only the one who sleeps, who passes his life sleeping, will not evolve (*end chorus*) and there is never an exclusive culture in the world.

Every culture is daughter of sub- and micro cultures,
every culture is a witness to a lived time,
is an experience,
every culture is a page of an encounter,
of a journey,
of a line of a poem,
of a relation.
Every culture is a witness to a lived time,
but not of the time to live,
because we cannot anticipate the culture that will come.

We know nothing about that, because certainly the Milanese culture of 3000 will have nothing to do with this. It is clear that it will draw some positive aspects that *we* would have left them but they will not have to move necessarily how we move [today] [...]

(Music ends as the woman with the accordion exits the stage.)

[Our immigrant friend] [...] will meet many people and offer his (her) articles [moving between two train stations and the city centre] [...] There are some times in which (s)he really succeeds in selling and making money, but there are also worse moments in which (s)he cannot even sell one product [...] (S)he will try in any case and in every way *not only* to sell his (her) articles but (s)he will also try to educate him (her)self through his (her) job, because there is not the necessary and sufficient time to go to school and learn the language. And so his (her) school becomes the street. And if his (her) school is the street, the people (s)he meets in the road become, so to speak, his (her) teachers. For this reason it is not uncommon to see one insisting to sell a CD. (S)he does not care [...] if the person in front of him (her) likes music or not [...] The important thing is that there is that dialogue which enables him (her) to understand the tenses, the accents, since Italian grammar is something that scares everyone [...]

So (s)he takes this direction and follows the crowd and finds him (her) self in front of the Castle.

(Lights go out, and when they come back on, only the right side of the stage is in the spotlight. Instead of the typewriter, there is the Sforza Castle. Mohamed/the tour guide and Cristina/the anthropologist are standing in front of it.)

MOHAMED/THE TOUR GUIDE:
But (s)he [our immigrant friend] is much more interested in the history of Piazza Mercanti than the one of the Castle. Because the Castle is yes, is a witness of the life and of the inheritance of famous personages like Leonardo and so forth, but it has little relevance for somebody who arrived in the city. Because in our part [of the world] a castle is lived like a fort, where inside [...] there are the very rich who have everything and outside the ones who are starving. That figure strikes his (her) sensibility and so (s)he confronts that reality differently [than, for example, a tourist] [...]

[Different European cultures and languages] are some of the things that he starts to understand by coming to the Castle, because it is a crossroads of cultures and traditions, of languages, because all the tourists, especially European, are fascinated by the Castle because [of] the name of [...] Leonardo [da Vinci] [...]

Except for our friend, who sees this Castle with diffidence [...] I personally entered only once in the Castle and I have been here for 6 years [...]

[Our friend] finds himself in two conditions that are often contradictory, right? Wanting to know and to learn, but also wanting to survive. One lives only through work/labour, so to combine the two things is not always easy [...] It always happens that [...] at a certain moment (s)he receives a phone call from a parent who is not well and (s)he feels on his shoulder the duty to respond to those needs. He is in a *precarious* situation and so what does he do? He makes violence to himself to help them. It could be that he is not well, but he will never tell them [...] also because he is cheated by the pictures that he sent to his country the day after he arrived in Milan. It is all an itinerary that will always shape his permanence in Italy, especially in the city of Milan, where there is a very strong and beautiful cultural inheritance [...] *(voices from backstage join Mohamed's voice, creating a chorus)* Culture is the only thing that is left to a person even when everything is taken away from him (her) *(end chorus).*

Take everything away from me, but not my culture [...]

(A church bell sounds.)

So, returning to the itinerary of our friend, following the footsteps of people who flow to the Cadorna Station[15] [...] there (s)he will see another dimension of the city, because for sure (s)he will have more refusals than [there are] days in the year[16] [...] (S)he sells but (s)he will also need information. Perhaps (s)he is hungry, or there is something (s)he needs, (s)he [...] uses the languages (s)he knows, but often (s)he gets ridiculous answers [...] The prejudice is very rooted from both sides [...] probably everyone has seen in him the potential *vuccumpra* [lit.: "wannabuy?" – a derogatory expression to refer to a street vendor, which is often used to designate North African immigrants in general] [...] Doing interculture would mean to challenge this taboo: look guys, it is not like that [...] every individual has his (her) own story, his (her) own experiences, we cannot use labels [...] then all Italians would be potential mafia members and all Africans potential vuccumpras. So in the end [...] he understands that his permanence in Italy will not always be a mousse au chocolat.

[...]

[Sometimes a person] finds him(her)self with our friend potential vuccumpra who wants to sell him (her) a CD. So (s)he refuses rudely. Yes, but in the end [...] (s)he walks ten steps and then comes back, and says: "Sorry, I did not mean it."

This establishes a connection, a relationship: [...] "Sorry, I did not mean it, this is not a day."

So the human reaction [of the CD vendor] would be "But why it is not a day? The sun has risen, and it will set in a little while. Why it is not a day?" [...]

"No, [...] it is not a day, because of a personal situation [...] because I am going through unhappy times."

So there one starts to do interculture, because in our part [of the world] it is always necessary to enjoy life – why? Because there is the knowledge that (*voices from backstage join Mohamed's voice, creating a chorus*) whatever we have, someone else desires it very much and lives it like a distant dream (*end chorus*)

[...] In Milan one would say *ciappo la vita come la ven* [I catch life as it comes – in Milanese dialect], which means to look at the positive aspects of life – which is what our immigrant friend was saying after all. And so, (*voices from backstage join Mohamed's voice, creating a chorus*) even if the world seems so old, the beginning of the future emerges always from the past (*end chorus*).

So this will bring our [presumably Italian-born] friend to reflect: [...] he who leaves his land, his country, his loved ones [...] perhaps coming here from a place where there is no winter, where it is always warm, and finds himself here in a cold, grey, country [...] he still finds the strength to smile at me and to tell me: come on, don't give up, you can do it [...] [This] will bring this other friend of his [the Italian-born passer-by] to confront his (her) problems.

All the encounters, all the relations, are born from this itinerary. That [Italian-born] friend the next day will spontaneously pass by […] only because (s)he will want to chat some more. There begins that voyage of encounter that sooner or later will cause him (her) to say: "But you, do you like doing this work?"

And the automatic response: "I have no choice."

So [the Italian-born resident] will […] look among his friends and […] relatives for anyone who could help […] The majority of us who have found a job have found it this way. They are relationships that last our whole lives. And most of us have a child whose Italian name, how to say it? The choice of that name is given from that experience. Thus, here [the street] is not only a crossroads of cultures, of people who come and go, here life stories are born. Here are born more positive aspects than negative […] As long as there is life, there is hope. The hope of each of us is to be able one day to give to this city what we have in our heart, that is our knowledge that (*voices from backstage join Mohamed's voice, creating a chorus*) whatever we have, someone else desires it very much, and lives it like a distant dream (*end chorus*).

Here ends our itinerary.

City Wars

And here ends our play. One of the many things that I learned from Mohamed Ba's performance is that perhaps we could come to appreciate public space as what could make it possible for us to switch between positions, to follow other people's itineraries, and to at least fleetingly participate in each other's lives and identities. This does not mean that we are all positioned equally in the city. Mohamed Ba is keenly aware of the constraints facing less-privileged residents of Milan and the deep inequalities that structure Milanese society. He talked very clearly of immigrants' experiences of crammed, unaffordable housing, of daily discrimination and labelling, of the hardships they encounter when work is slow. He explained that it is very difficult to find a job other than selling in the streets, and that it is necessary to know somebody even for being considered for employment. He showed how vendors' experience of the city is marked by tiring daily journeys, especially in winter when it is very cold and they have to stay long hours outdoors.

Yet, his narration and journey through the city highlight both the limitations and possibilities of sociality in public space. It poses the question: what if? What if differently positioned people could encounter each other in public space? What if they/we could engage in "interculture"?

The performative structure of Mohamed Ba's talk is important for the creation of this imaginary. By using streets and piazzas as stages, he establishes a complex correspondence between speaking and moving, walking and telling, or the itinerary and the story. One is created by the other. By tracing parallel journeys and maps through the city, he crafts an open text, where the listener can move and imagine different voices, experiences, and positionalities. In this way, I felt that Mohamed Ba called me to witness the very possibility and power of the imagination. Fabian (1990), as well as Thrift point out that performance can "expand the existing pool of alternatives" (2003: 2021) and the repertoire of ideas, dreams, and memories available to us.

This is no small feat, if we consider, as Dolan does, that imagination might be the necessary bridge towards utopia. Theatre is a public practice through which the "field of the possible is [...] opened beyond that of the actual" (Ricoeur, 1991, quoted in Dolan, 2005: 89). Glancing at possible alternatives to the status quo helps us see utopia not at a fixed state that can never be reached, but as a process, a desire, and an affect that emerges in particular moments in our daily lives.

In the context of urban space, the experience of this imagination might encourage us to search for those moments in which the public space that is created through extraordinary, everyday encounters – as the one enacted by Mohamed in his concluding vignette – allows for connections and alliances between residents and generates counterarguments to powerful discourses. The word "enacted" is here crucial. By performing streets and piazzas as complex journeys of hope, sociality, and discovery, Mohamed Ba fashions a place and time in which people and positionalities that might otherwise not interact with one another can confront and add to each other. To borrow Mohamed Ba's beautiful metaphors, he offers an ephemeral moment[17] in which they can make music together or try on new dresses, while always being aware of the gap between notes or of the subtle distance between bodies and garments.

Placed somewhere between a formal theatre play and an unstructured, daily walk in the city, Mohamed Ba's intervention could be a provocative companion to Dolan's questions: "How can performance model civic engagement in participatory democracy? How might performance let us rehearse truly democratic public practices through a kind of social mimesis? That is, instead of art imitating life, how might we bend life to imitate theater, with its necessity for attentive listening, for dialogic reciprocity, for the company (and kindness) of strangers?" (2005: 90). Mohamed Ba's walk suggests that streets and piazzas can, at times, provide an answer to these questions. The form of public space modelled by Mohamed Ba, "with its necessity for attentive listening, for dialogic reciprocity, for the company (and kindness) of strangers," and for vision as an embodied witnessing and a critical act of representation, can at times "model civic engagement in participatory democracy" (ibid.). I find this idea especially precious in a city where history has been largely fought out in the streets. For one, Mohamed Ba's words alert us to the acts of performance that often characterize these conflicts and the making of their memories. For another, they remind us that while this performativity helps constitute the urban terrain as an unresolved site of struggle, it can also encourage the spatial and social imagination that creates public space as an extraordinary daily possibility.

Mohamed Ba's narration might then be helpful in illuminating a series of events – three rallies and two epitaph installations – that happened in Milan almost exactly one year after the walking tour represented above. Considering the demonstrations and placards after having walked with Mohamed Ba, we can see that they, too, albeit in a different way, use piazzas and streets as stages to tell the history of Milan. In turn, the events of March 2006 are evidence, once more, that Mohamed Ba's words are not simply a story about Milan. Because streets and piazzas are crucial political sites – a medium and result of ongoing struggles – his journey is an act of engagement that helps shape the city itself, its landscapes, and its identities.

On Saturday morning, 11 March 2006, a demonstration took place in one of Milan's major shopping avenues, Corso Buenos Aires. It was unauthorized by the city and organized, so it was reported, by people associated with the Milanese Social Centres. This rally was a counterdemonstration to a fascist parade planned by Fiamma Tricolore (Italy's far-right political party) for that afternoon and authorized by the municipality.

For reasons and mechanisms unknown, during the morning rally, a group of young demonstrators burned cars, smashed shop windows, and injured a group of policemen. About twenty to thirty people were arrested for this action. In an additional surprising turn, bystanders charged the demonstrators who had been held up by the police, who then found themselves in the ironic position of having to defend their own captives. This is how Ansa Italy reported the events:

The demonstration of the Social Centres, at noon, started within a very tense climate. About 200 youths, many of them with helmets, with their faces covered with balaclavas/masks, with wooden sticks in their hands, marched from the Lima Piazza to Porta Venezia [also a piazza], where there was a large anti-riot police force [waiting] for the non-authorized rally.

There was a dense throwing of stones and firecrackers [...] against the police. The demonstrators from the Social Centres also set wood bundles and garbage cans on fire close to the *caselli* [two little buildings that were once part of the historic city wall] of Porta Venezia. The police [lit.: "the forces of public order"] threw tear gas while the firefighters succeeded in extinguishing a fire from a scooter and a newspaper stand; the firefighters could not, however, come close to some cars which were also engulfed from the flames [...] The shocked mass of bystanders turned their anger against the 2–300 demonstrators who had unleashed this disorder, when the latter were held by the police. Just barely, in fact, could the agents save them [the demonstrators they had arrested] from a real lynching: large groups of people were beating them, shouting "Kill them!" while the police were hardly managing to load them into the vans.

A store, close to the corner of Corso Buenos Aires and Viale Regina Giovanna, went up in flames, like two cars that were closeby [...] The flames were cordoned off by firefighters, who also evacuated many apartments of the building, invaded by the smoke. (Ansa News, 2006)

Meanwhile, in the afternoon, the authorized right-wing manifestation took place undisturbed, following the same route as the morning rally. A couple of hundred people, escorted by police, paraded invoking the name of Mussolini, shouting fascist slogans, displaying Italian and Fiamma Tricolore party flags, and performing the Roman salute.

These two rallies and their dynamic sparked yet another public display. The following Thursday, 16 March, the owners of businesses in the area (Corso Buenos Aires) organized a torchlight demonstration against the violence of the Saturday morning rally – thus becoming a rally against a rally against a rally. According to *Corriere della Sera*, about five thousand people participated, including representatives of several political parties (but carrying no political party flags or signs):

The torchlight rally started a few minutes after 8 p.m. [...] Many Milanese and Italian flags are being carried by the demonstrators in Corso Buenos

Aires, where [...] the windows and awnings of shops have remained lit in a sign of protest [...] A plaquard with the writing "the city that lives wins" opens [i.e., is carried by the front row of] the rally. ("Prodi e Fassino Disertano la Fiaccolata," 2006)

Reading the above descriptions from the newspapers (unfortunately, I was not in Italy at the time, so I could not be part of those events), just after relistening to Mohamed Ba's commentary, I was struck by how these reports themselves sounded like stage directions. I could not help but rethink the comment by Pier Paolo Pasolini, one of Italy's major theatre writers and film directors: "The archetype of the theatre occurs before our eyes every day in the street, at home, in public meeting places, etc. In this sense, social reality is itself a performance that is not entirely unaware of its being such and has, therefore, its own code" (1983, quoted in Van Watson, 1989: 23). In Corso Buenos Aires, the burning cars in the morning, the Roman greetings in the afternoon, and the torches of the evening (to name just a few) responded to each other as they created a very complex choreography. By this I do not mean that it was not serious business. On the contrary, the very chain effect of these public displays confirms, once more, how the streets can be, and often are, highly contested and important locales for cultural, economic, and ideological warfare and for the negotiation of social realities. Although this can be said for many cities around the world, it is especially true for Milan. Demonstrations and public conflicts, be they large or small, quiet or intense, are not an unusual sight in the city – now, as well as in the past. After all, the "five days" of Milan that Mohamed Ba was talking about earlier refer to an 1848 battle (more precisely from 18 to 22 March of that year) in which the people of Milan built barricades in streets and piazzas to free their city from Austrian rule.

Fascist and left-wing forces especially have historically engaged in conflicts that have marked the urban environment and the memory of city places (see, e.g., Foot, 2001: 14). It is perhaps telling in this respect that the two women over 80 I talked to during my research remembered the end of Fascism as that day in April 1945 in which Mussolini's and his fiancé's bodies were hung in one of the central piazzas of Milan, the Piazzale Loreto. Coincidentally, and perhaps ironically, this very same piazza is one of the end points of Corso Buenos Aires and thus was one of the locations for the March demonstrations.

Mohamed Ba's tour suggests that one of the reasons why streets and piazzas are such an important medium for the negotiation of social

realities and the constitution of political identities might be because
they are open to the very possibility of performance and performativity
(see also Freeman, 2001; Fleetwood, 2004): people routinely appropriate streets and piazzas to represent, embody, and reinterpret identities
and spaces, thus participating in ongoing debates on public histories
and meanings. This, of course, can be as easily progressive as it can be
conservative, repressive, or totalitarian (as the events in Corso Buenos
Aires remind us).

In using the term "performance" in this context, I would like here to
expand its meaning from a more specific and formal theatrical genre to a
more metaphorical understanding of the term (without losing sight of the
porous boundary between the two). A large body of literature has been
focusing on performance in the context of everyday life to bring attention
to speech and language "as social action" (Bauman and Briggs, 1990: 62).
Rather than isolating and privileging the content of what people say, a
performative framework helps us consider "to [and with] whom, when,
how, and why" (Fabian, 1990: 8) people engage in acts of telling, retelling,
listening, and remembering. Scholars who attend to how people enact
identities, spaces, and social relations seek in this way to analyse both
the social constructed-ness of the world and the ways in which people
add new twists and meanings to discourses, systems, and situations. As
Gregson and Rose suggest, the notion of performativity helps us look at
the "creativity ... and uncertainty" of daily life (2000: 434). This creativity,
as noted by Thrift (2003), is politically significant, as it includes the ways
in which people often, so to speak, interfere with the script.

Public spaces work well as informal, metaphorical stages for a variety
of reasons. As Nicole Fleetwood points out regarding public transit,
public spaces present a mix of anonymity and spectatorship: although
to a lesser degree than when on board buses, by being together in busy
streets and piazzas, people are "forced to bear witness" (2004: 37) to
others' appearances, words, and movements. There is, then, no place
outside of the stage. As Mohamed Ba exemplifies, public spaces are
replete with layers of meanings, events, and memories. The traces of
these stories can be harnessed to perform other ones and in so doing
to reinterpret the past and the future. Public space serves as one of the
arenas where our iterative gestures, poses, words, and movements crys-
tallize who we are both for ourselves and for others; see also Guano
(2007) and Del Negro (2004).

Mohamed Ba's walk suggests that we look at public space as an
assembly of stages framed by particular publics, interests, events, and

stories. Public memory is particularly apt at being displayed, performed, or obliterated in public urban locales. Signs, monuments, expositions, parades, and murals are just some examples of the many ways in which collective recollections are made, maintained, and reframed (Burk, 2010). Yet, memory, like vision, is not simple, transparent, or fixed. Remembering Mohamed Ba's play with what can and cannot be seen, we may ask: how does the violence of burning cars become much more visible and understandable than the violence of enacting a certain type of remembrance? For whom and in which "regime of the visual" (Guano, 2002: 305) is one clearer than the other?

Interestingly, the Social Centre Leoncavallo responded to the accusation that youth from the Social Centres participated violently in the morning rally that March Saturday by reminding the city of the tragic disappearances and forgettings that mark the history of Milan. These include the murder of Dax, one of the youth from the Social Centres whose 2003 death anniversary falls on 16 March, and the killing of Fausto and Iaio on 18 March 1978 (see chapter 4). The latter case has been recently archived without finding out who shot the two young members of the Leoncavallo, but many believe that it was "a political murder against the left" (www.faustoeiaio.org) and that it happened because Fausto and Iaio were involved in an anti-drug campaign (see Membretti, 2003: 92). According to the Leoncavallo, it is important to understand and evaluate the current events as part of wider conflicts between fascist and progressive forces that have involved urban spaces in Milan at least since the late 1960s and that have direct linkages to political elections and governance. (That the rallies happened some weeks before the municipal elections in which Moratti became mayor adds another dimension to all this.)

Mohamed Ba's words and the comments from Leoncavallo help us understand some of the links between space and memory that emerged during that March week in 2006. The situation, indeed, became a conflict about the public memory of violence. On 18 March, just two days after the third rally, in a night blitz, municipal officers replaced an epitaph in Piazza Fontana, the site of the 1969 bombing by as yet unknown perpetrators. The sign, which commemorated Giuseppe Pinelli, the anarchist who lost his life in police custody in conjunction with the bombing, suddenly declared that Pinelli "died" instead of "was killed." A few days later, on 23 March, the anarchist association Ponte Della Ghisolfa replaced the sign that had been in Piazza Fontana since 1978. Figure 34 shows how the piazza and the two signs looked in March 2006 (and still looked in 2011).

Figure 34. Commemorating Giuseppe Pinelli (photograph by Enrica Sacconi, March 2006). The sign *to the left* reads: "To Giuseppe Pinelli anarchist railway worker killed innocent in the rooms of the police of Milan on 16/12/1969. The students and the democratic people of Milan." The sign *to the right* reads: "Municipality of Milan. To Giuseppe Pinelli anarchist railway worker innocent who died tragically in the rooms of the police of Milan on 15/12/1969."

Although Mohamed Ba's tour seems at first sight "just a story," and these rallies and events "real war," their juxtaposition highlights the power of performance and the performative aspect of power. Several authors come to mind here. Diana Taylor, in her analysis of the "Dirty War" in Argentina in 1976–83, argues that the dictatorship used spectacles to create consensus and to consolidate the power of the regime: these public enactments included the embodiment of the state by the three national leaders, the "staging [of] order" (1997: 67) and masculinity by the military, and the representation of the hero as a "lone soldier" fighting against the forces of death, femininity, and moral decline (73ff). Guano suggests that performances not only can legitimate the ruling powers, but also undermine them. Discussing the fall of President Menem in the Argentina of the 1990s, she shows how teachers on strike and the population supporting them used rallies and demonstrations to cast themselves as the protagonists of their country and of political dissent; in this way, they were able to displace the government from centre stage into a role of spectator and thus to destabilize the "modalities of seeing, displaying, watching, and being seen" (2002: 303) that were so central to its politics.

Like Mohamed Ba shows in his narration, both Guano and Taylor argue that an important way in which performances sustain or challenge structures of power is by reframing and reorganizing visibilities and invisibilities. The dictatorship in Argentina, for example, was based on an "unequal visual economy it established with the public" (Taylor, 1997: 71), in which leaders were "on display" while refusing to "return the look" (ibid.). Indeed, the role of vision in public space, and in creating political identities, brings Guano to suggest that we look at public space as a site of performative praxis, and the public sphere as the result of "largely performative arenas where agentive participation can be established through a visual economy of critical spectatorship and performative action" (2002: 306). Demonstrations like the ones that took place in Milan in March 2006 are perfect examples because, in a way, they are intensely theatrical. People taking part in them show themselves as political and social actors, and participate in the negotiation and representation of memory and social realities. Mohamed Ba's walk, however, points out that performative practices and commentaries matter not only amid crisis and dramatic events, but also in people's everyday lives, as they imagine, remember, narrate, move through, and use city spaces.

I juxtaposed Mohamed Ba's tale with the March 2006 rallies because, as Thrift reminds us, often performances and poetics are thought of as apolitical (2003: 2021). As Thrift notes, they are often seen as "arty stuff" (ibid.) detached from social engagement. Similarly, a focus on vision might be seen as irrelevant when it comes to understanding and intervening in Milan's "real" world. After all – some may say – what difference does it make to the everyday conflicts and debates in the city, if we imagine with Mohamed Ba different musicians playing together or travel with Bellovoso and his councillors? Here I would like to suggest otherwise. Look at what happens with the streets: they are seething with mysteries of which we are always a part. They demand of us that we look twice, because, just like the two signs for Pinelli, public spaces harbour many kinds of truth. Or, again, look at the placards in Piazza Fontana. In a way, these signs do not just commemorate Pinelli. By telling a story twice, but differently in each case, these signs create a place for an audience as active onlookers who cannot but question "what really happened" in 1969. Like the demonstrations, they are not just responses to facts or the recalling of something that has taken place. Rather, they are ways of making history matter and of constituting public space as a very medium and result of those struggles. These signs, moreover, are themselves strikingly similar to ghosts – companions, perhaps, to dismissed areas and saints with too many arms. Their significance as part of a controversy works in the same way as a "spectre": "it *begins by coming back*" (Taylor, 1997: 30, quoting Derrida, 1994; emphasis in original).[18] Like ghosts, these signs tell us that seeing is not sufficient. Indeed, they literally *show* us that looking is not enough. By representing a contradictory message, they seem to mimic the "seeing double" of when we do not see very well. By enacting for us a double truth and double vision, these plaques suggest that all urban landscapes might be like Piazza Fontana: unruly characters in a dynamic relationship with the play, through which they are made, undone, unravelled, reorganized, and reinterpreted.

Considering just how complex, contested, absurd, surprising, and even outright violent the life of public spaces can be, what other tools do we have but performance, haunting stories, and alternative perspectives to comment on it and imagine better possibilities? What other strategy but to engage people's imagination, "affect," and the "immediacy of the now" (Thrift, 2003: 2020)? By locating hopes, creative intervention, and utopian moments in everyday encounters and common public spaces, Mohamed Ba's performative walk opens a dynamic space we could move, live, and learn in.

Conclusion: Into the Future

Every year in Milan the association Esterni organizes a Public Design Festival. The goal of this event is to promote public space as a key site for the production of culture and society. Central to Esterni's work is the ideal of the piazza as a ground for interaction, a place where people can actively participate to make "this thing called 'city'" (Zenobia presentation, 9 March 2005). Following this mandate, on 18 April 2009, the organizers of the Festival occupied ten car parking spaces on the streets of Milan and replaced them with interactive "stations" – like a treehouse for children, an eight-person bicycle to produce energy for playing music in the street, and a swinging, communal bench.

One of the ten sites consisted of a public feet-washing station. A line of chairs was set up along the street, and each chair was equipped with a basin, a jug full of water, a towel, and two flacons of perfumed soap. In the moment I arrived on the scene, all but one of the chairs was occupied. People were washing and drying the feet of those sitting on the chairs, as a group of twenty or thirty bystanders looked on, laughed, chatted, and drank wine offered by two of the organizers who were circulating in the back. In a city where most of the population is familiar with Christian doctrines, the symbolic (biblical) meaning of washing other people's feet – representing humility, equality, and service to others – could not have been lost on the mass of onlookers. It was an intriguing contrast to the traffic of the cars rushing nearby and to the business-minded anomie and individuality that Milan is known for. Before I even realized it, my two-year-old daughter had jumped onto the remaining empty chair, delighted at the thought of having her feet pampered, and a nearby woman had started immediately to pour water in her basin, ready for the task ahead.

This station was fascinating to me because it succeeded, more than any other in the Festival, in showing that people make public space and do not just take part in it. To say it in Don Felice's words, during the feet washing the "asphalt was warm," and the street alive, filled with a sense of community and with the excitement of engaging with a surprising idea. However, as the people got off their chairs and moved somewhere else, the "sand dune" moved, too; when I returned to that street an hour later, the empty parking space dotted with basins, towels, and jugs was just that: a row of empty chairs.

In this book, I have attended to how public space as a concept and a dynamic dimension of the social comes alive in the comments, narratives, and performative everyday engagements of particular people and groups. The examples I discussed show that different Milanese construct particular understandings of public space for various purposes and with various consequences. Some resonate with powerful discourses, some are more marginal and can work like "a hair in the flour" (Tsing, 2005: 205) of official narratives.

The VivereMilano forum evoked an understanding of public space as an agora for the city, which both celebrated an ideal role of the piazza in urban affairs and effectively restricted the kinds of conversations, performances, and subjectivities that were invited to be part of it. The Social Centres also use piazzas and streets to voice dissents; however, their critique of the ruling power relations that shape the urban territory brings them to build spaces of resistance and political opposition in the city's interstices. Rather than simple locations, these spaces are part of a project of critical geohistory and are best understood as processes aimed at revaluing memory and reformulating history.

When commenting on urban spaces – including parishes and community organizations, employers' homes, parks, streets, and piazzas – the Milanese I talked to who had migrated to Italy from other countries described coexisting experiences of constraint and discrimination, as well as opportunities for critical interventions and for sociality. Their relationship with Milanese locales needs to be seen in the context of dominant discourses on immigration. Here "public space," as a fluid yet significant referent, becomes a medium for asserting essentialist associations between identity and the city, which forget that if space is always constructed and in flux, so is also cultural identity and the very experience of "strangeness."

These different orientations, enactments, and interpretations of space reveal patterns of urban inequality, contested histories, unequal access to city spaces, and everyday conflicts and alliances between inhabitants, in a time characterized by rapid urban change and neoliberalization. This, however, rather than affirming a definite social category we can call "public space," troubles the very term and muddles its contours. The more "public space" is used as a referent to comment on these issues, the more it emerges as a shifting concept that eludes definition. Importantly, it is this very instability and difference that enables public space to become a discursive, embodied, and performative realm where different inhabitants debate their and other residents' places in the city.

Vision as an idiom and a practice and (in)visibilities as avenues and results of social processes are integral parts of these practices of place-making. People become visible and/or unseen in the city in many different ways, places, and times, depending on their social positions in Milan and in wider economic, cultural, and political contexts. In turn, their acts of looking, recognizing, and making themselves seen (or not) reflect and negotiate social relations and structures of power. In Milan today, wide-ranging social shifts (from the growing of social divides, to urban renewal, a widening multiculturalism, and neoliberal restructuring) are consented to, challenged, and elaborated also through ways of seeing, forgetting to see, showing, concealing, and attuning oneself to certain landscapes.

Nuanced practices of seeing and being seen illuminate the working of powerful categories and processes that shape spaces and identities, as well as people's interpretations of and interventions in those social forces. The example of the promenade, the *struscio* – waning as a social practice happening in specific times and places but enduring as a generalized mode of showing and perceiving self and others – is instructive in this respect. It shows how a simple walk through the city involves attention to people and places in which vision helps to situate bodies in landscapes and negotiate the relations of inhabitants to one another.

Participating in piazzas and streets in Milan is a particular visual experience where it matters what and who one sees, and where. In the neighbourhood where I resided, for example, the striking passage from the elitist fashion houses to the cheap department stores, from the vacant dismissed areas (*aree dismesse*) to the trendy cafés and the run-down social housing is not just a matter of very different backdrops that accompany one's walk. They shape in important ways one's experience of being in space and participate in the making of streets and plazas in a complex process of mirroring, interpellating, questioning, distancing, and inviting. In this charged and dynamic visual context, subjects are always and inevitably linked to their surroundings, and each inhabitant is always marked by his or her own gaze towards fellow residents, as well as by the looks of others.

In examining public spaces and their visual lives, my intention has been to privilege my interlocutors' explanatory frameworks, modes of inquiry, and insights that emerged during my fieldwork. To name just a few, the activists involved with Social Centres (*centri sociali*) such as Casa Loca and the Stecca degli Artigiani suggested that I look at spatial phantoms. Mohamed Ba taught me a model for public space, according to which intersecting stories and itineraries enable the depth of history and the power of imagination necessary for engaging with different voices and perspectives. Francesca helped me understand the shifting boundaries between public and private spaces. And Eliza helped me to rethink the politics of time. These "local theories" are important because they are avenues for commentaries, critiques, and insights regarding everyday life in Milan, and its social, economic, political, and cultural conditions.

Milan today is a city facing tremendous changes. Many commentators see Milan as having lost its vocation as a centre for work, solidarity, and innovation; they see it as being in some respects a failed metropolis, lacking leadership and a long-term strategy (Goldstein and Bonfantini, 2007). This pessimism has been deepened in the past two decades by enduring social inequalities, widespread racism against immigrants, the growing "precariousness" of life, and by a general disillusionment with the political and economic situation of the country (Molé, 2010; Guano, 2010).

All of these tensions come to the fore spatially, in everyday trajectories and encounters. Luca Doninelli, for example, poignantly describes how Milan "awakens twice" every morning: first, with the rising of those who inhabit its margins and interstices and then again when the "houses, the coffee makers, the microwaves, [...] the traffic," the stores, and the more well-off inhabitants start their day (2010: 161 and 162). The places and subjects of the first awakening – the mass of people who, from the periphery of Milan, take the very first subway and bus rides to the city at 5:30 a.m. – are most often invisible to the rest of the city, yet the people of the "later shift" will live their day deeply connected and in many ways intimately dependent on the former population. In Doninelli's Milan, even the simple fact of waking up is a complex encounter between different realities, with the very streets and piazzas bearing the imprints of these negotiations (see also Novak and Andriola, 2008).

Everyday occurrences during my fieldwork echoed this sense of fragmentation. Consider this, for example:

As I approach the bus stop, I am struck by the sight of two people sitting on a small bench. One of them is young, between 30 and 40 years old. He is well dressed: dark pants, an ironed shirt, and a tie. He is holding a briefcase, and it looks like he is on a break from work. Just beside him on the bench, is an older man with white hair and in simple attire. He is playing an accordion, and he has placed a hat in front of him hoping for the generosity of passers-by. The two men are sitting so close to each other that they could be friends, yet neither of them seems to even notice his companion. This scene, and its incongruity, reminds me of some of the contrasts in contemporary Milan: the wealthy living alongside the poor; the well-dressed, busy professionals walking past the low-paid, temporary labourers. What is striking – as in the case of the two men at this bus stop – is that they share the same spaces yet seem great distances apart. Often, like for this bench, it becomes also a difference of generations. Among the growing poor in Milan are elderly people on meagre pensions. (20 May 2011)

These disjunctures between inhabitants are brought into sharp relief by contemporary redevelopment projects unfolding in Milan. These changes, happening as I write, will shape not only the lives of particular Milanese public spaces, but also the broader ways in which they are understood, lived, debated, and enacted by its residents.

It is important to note here that the public spaces planned by the developers of the new neighbourhoods under construction are one of the central selling points in their promotion. In the Garibaldi Repubblica area, for example, where a part of the Isola neighbourhood is located, a park constitutes the centre of the new developments. Pedestrian promenades are an important component of the planned streets. Cultural centres and the Museum of Fashion are being hailed as valuable additions to the urban life of the area. And the new regional administration building offers a covered, publicly accessible piazza as an entry point for its employees and visitors. The municipality of Milan, in fact, describes it as an extensive public space: "a piece of city [...] that can be inhabited, traversed, visited, and used by the citizens/city inhabitants (*cittadini*)" (Urban Centre, 2009). Similarly, in City Life – a vast, half-completed project in a less central part of Milan – green areas are described as the "nucleus" of the whole neighbourhood, and it is around this public space that the residential buildings will be located (ibid.). Moreover, according to the municipality's description, "the surface of the new neighbourhood will be entirely pedestrian, characterized by streets, piazzas, and paths" (ibid.).

Although it is too early to know how these new spaces will be used and by whom, the realization of new public spaces in these and other urban projects is inevitably entangled in the issues and dynamics I discuss throughout this book – such as the notion of security, questions of affordability and inequality, the cultural and economic role of fashion, and the right to city spaces by different social and cultural groups. For example, when I interviewed in 2011 a real estate agent who had been working for many years in the area that now includes City Life, he told me that although everybody was happy that new green spaces were being created, many residents were already worried about who will be using the park of the new development and whether this will make the neighbourhood unsafe. He pointed out that this might generate conflicts especially considering that the new residential towers are extremely expensive and cater to the affluent.

And a recent study on the almost completed project of Santa Giulia, a new neighbourhood established southeast of Milan, at a distance from the centre, suggests that its new public spaces, rather than being lively areas of sociability as planned, have become more akin to barren areas between the different sections of the residential complexes. According to Paola Savoldi, the parks and streets that were supposed to generate and strengthen a sense of community in this brand new neighbourhood are mostly empty or have become mere passageways – she poignantly describes them as "precipices," places that need to be "circumnavigated" (2010: 61). Part of the reason for this is that they were designed to make visible and control less-affluent residents or passers-by in the name of security. Because of their placement, moreover, between more- and less-privileged parts of the wider area, rather than counteracting class differences, they have made these very divisions more apparent.

What is at stake are not only particular parks and piazzas (although who will use them and how is undoubtedly important), but also wider ideas about who belongs to the city and how urban space should be assigned, used, and developed, and, indeed, even the idea of what the "public city" as a model might entail. The latter, suggests LaboratorioCittàPubblica (2009) can be understood as one component of a set of interlocking urban resources and common goods (including social housing, publicly owned and distributed water, inclusive welfare, and active Social Centres) that form the basis for a more equitable city – in opposition to a metropolis increasingly governed by market forces and autonomous, entrepreneurial subjects.

Although an examination of urban regeneration in Milan exceeds the scope of this study, I would like to briefly address this issue here because attending ethnographically to public space as a point of departure can prove an especially useful strategy to trace allied or divergent – at times, even incommensurable – perspectives and understandings of these urban changes. Current construction zones, I would argue, emerge as yet another kind of public space. They are conflictual sites of public engagement, tools of the imagination, and places where different inhabitants negotiate stories and relations, and where hopes and interpretations are clung to. Especially I am interested in how redevelopment projects activate a particular understanding of the future that situates certain ideas, identities, and positions into its grasp, as it casts others obliquely.

Acts of looking and depicting are, once again, part of these processes. The emergence of buildings on the urban territory is not just a matter of adding new elements to an existing terrain: it also trains our vision, encouraging us to look at the city in a different way. Of interest in this context is a promotional video for the new redevelopment processes in Milan created by the municipality (Urban Centre, 2008) – consisting of many minutes of dramatic music accompanying the simulated emergence of new skyscrapers all over the city. The video provides only minimal information to the viewer. The main message, as stated at the beginning of the video, is to simply show that the skyline of Milan is changing and to inform people that now Milan is a "city that grows," in the double sense of the expression: its buildings are reaching new heights, even becoming skyscrapers, and the city is creating new infrastructure as well as business and investment opportunities.

The film, rather than discussing a process, simply confronts us with a view, a dynamic panorama of Milan. As we watch from above over the entire area of the metropolis, we see the main buildings of the redevelopment areas grow, one after another. As each is drawn with a glowing neon line over the existing city outline, its height and name appears on the screen. Here we do not just see a landscape. That landscape becomes our way of seeing, and a way of imagining the future; it is, at the same time, a material reality and a particular technique of looking (Grasseni,

2011; Blomley, 1998). As Bricocoli and Savoldi (2010: 197) argue, it is a perspective that privileges "a single dimension, the vertical one, the one of heights and rises"; here, the "often evoked skyline [...] is the space of the staging of the city, its mise-en-scène [...] corresponding [...] to a government that wishes to leave a sign."

To conclude, I invite you to follow me in one last walking tour, in the Isola neighbourhood, as it sits uncomfortably on the edge of becoming. More than a closing, however, this is meant to be an opening for further questions, connections, and inquiries – the beginning of many other itineraries.

Questions and Footprints

Early on a Sunday in May I walk in the Isola, along the fence of the Garibaldi Repubblica construction zone, looking for the so-called Temporary Stecca. After the activist hub and Community Centre the Stecca Degli Artigiani was demolished in 2007 (see chapter 6), several of the organizations once housed in the disused but occupied factory took refuge in a building nearby, waiting to find another home. They had moved to a street called Vicolo (lit., small street) Castilia, and although I did not know exactly where it was located, I was determined to find it and visit it.

So, here I am, walking along a wall bordering a dusty void. After a few minutes of walking, I find myself in Castilia Street, wondering if I am going in the right direction: perhaps the Vicolo with the same name is close by? I come to a green fence dividing the sidewalk from an open space, behind which I see cranes rising towards the cloudy sky. Suddenly, I am struck by the feeling of having been here before, even if I do not recognize my surroundings. I realize this must be where the old Stecca and its garden had been – I remember sitting in the grass amid the stalls of the farmers' market, and the crowds pouring through the narrow building covered with posters and resounding with music. But these are only memories; today everything is quiet.

I see two elderly women looking at a sign on the wall opposite the green fence. The notice they are reading details the work in progress of a small new building in this street, which looks like a glass cube. Pointing to the empty terrain over the green fence, I ask them, "Was it here that there was a garden, and the Stecca degli Artigiani?" The two women say, "Yes," and they add, "The new Stecca is this square, shiny house here." They gesture to the small cube, and explain that although the building is empty, it is completed and ready to be used. They tell me to "go see it from the other side, [because] it is more beautiful seen from the side than from the front." "Go and see it!" they insist,

> Go inside the children's playground. There is a restaurant, from the side of the restaurant you can see it well, since it is more beautiful from the side.

They say this several times, as if the main point of the building were to offer a pleasant view to passers-by. Then, they turn and notice a man with a heavy camera hanging from his neck. "Let's ask him," says one of the women, because "he knows." The man confirms that this is the New Stecca, all ready, yet empty. None of us knows why. The women and the man with a camera go on their way, and I remain in the street still looking for my destination.

Left with more questions than answers (why does the man with the camera know? who is the cube for?) I recall a promotional film I had recently watched about this area, made by the developers of the new neighbourhood, called Porta Nuova. In the video the voice of a young girl tells the history of this part of the city:

> My grandmother said that trains used to come past here. Choo! Choo! Lots of people coming and going, she said. Then the city got bigger and bigger; the trains moved over there.
> [...]
> Then they build the subway [...]
> Today also, things continue to change. Because Milan never stops.

The video continues with a series of short comments by people involved in the project and by one community resident. After they have spoken, and as the camera surveys the many amenities and the general landscape of the redeveloped area – as it is supposed to appear when it is completed – the girl continues her talk:

> Do you recognize this? This big area is now called Porta Nuova.
> It revolutionizes the centre because it creates new links to all the neighbourhoods around it.
> There are the regional and city council buildings, there are offices, a large hotel, shops, lots of houses, and plenty of public and private parking spaces.
> I live in an apartment overlooking the Giardini of Porta Nuova, one of Milan's largest parks. It is such a perfect place for walking or cycling [...]

The cars go underneath the round piazza so that the Corso Como [Avenue] and the Isola area are finally connected via the park.

If I have to leave Porta Nuova I have many transport choices at my doorstep [...] It's convenient to have them closeby, and above the silence of the park.

There are also five cultural centres in Porta Nuova, with everything from fashion to arts and crafts. Porta Nuova will bring new vitality to one of Europe's most extraordinary historic cities.

The girl's voice, moving from the past to the present day and then to future changes, is accompanied by suggestive images and music. The first part of her tale, referring to her grandmother, is paired with old newscasts of Milan, showing busy commuters in black and white, the coming and going of trains and the subway, and the demolition of old buildings. The description of her life in Porta Nuova is echoed on the screen with views of the simulated new neighbourhood, and it is narrated while electronic lines form on the screen, drawing new buildings within the existing territory.

Before and during the second part of the girl's speech, the video brings us in an imaginary flight over the area. We enjoy ample vistas from above, as the camera lens moves backward, forward, around, and we feel ourselves alternatively swooping down and gaining elevation. Towards the end of the girl's commentary, this movement takes on a decidedly forward motion. It is as if she herself was flying rapidly through the space, between houses, above the buildings, and through the park. At times, she is at a person's height, to simulate inhabiting the Porta Nuova complex and using its paths; at other times, she passes high overhead to amplify its vastness and expanse.[1]

Starting out with a decidedly childish voice and assuming a slightly older tone in the later section, the girl is not simply showing us the neighbourhood – she is moving through time. This is accomplished not only by her words, but also by the itinerary she embodies. In her flight through the city, as she rapidly overtakes the stationary silhouettes of persons that dot the computer-simulated landscape, she is the one who can move beyond and ahead. This capacity for reaching and overtaking is rooted in her understanding the need for transformation: everything continuously changes, so why not this area? This results in her, literally, inhabiting the future. She declares, in the present tense, that she lives in the apartment building by the park and uses the subway below her – neither of which has been built yet.

As the one who remembers and knows how to interpret the past yet is flexible and dynamic enough to reach the future, the girl connects different eras. This trope is paralleled by another key metaphor: uniting the fragmented space of the neighbourhood. In the video, the interviewed men explain that this zone of Milan "has been isolated," that it represents "a deep wound in the fabric of the city," and that we need to "recompose" its "divided" parts. It is in this light that we should consider the girl's concluding remarks: "Corso Como and the Isola are finally connected."

As Erik Harms argues in relation to Ho Chi Min City in Vietnam, urban spaces that are at the edge of becoming take on characteristics that are at the same time spatial and temporal: "In terms of urbanization, the idea of time as a forward march of progress seems to emerge from the idea of spatial expansion occurring over time. That is, the march of time seems to take on material reality because it builds from a spatial metaphor of urban expansion that seems to objectively mark the progress of time" (2010: 114). As Harms reminds us, the future is never an innocent category. The way it is invoked, and who contributes to its imagining, have important consequences for how the past is remembered and the present constructed. In Vietnam, notions and discourses about time enable people to describe places, such as urbanized centres and rural hinterlands, and to rank them according to socially constructed ideas of modernity and progress. The countryside and the village come to symbolize backwardness and tradition, as opposed to the future-oriented, modernized, urban core. Similarly, in Milan, the girl's tale, tracing and accomplishing the passing of time, unites the future and the past by mapping it onto the landscape. In this way, space itself becomes an embodiment of the forward movement of eras.

Remembering the video now, as I stand in front of one of the fences of the massive construction zone, I am struck by how different the girl's view and movement is from mine. Why does she know so well where she is going while my walk hinges on encounters, snippets of conversations, and interrogations? This is not necessarily a difference in knowledge. After all, I grew up and lived for twenty years a ten-minute walk from here. I can also refer to *my* grandmother when recounting changes in the urban landscape. What I want to point out here is that there might be advantages to a mode of knowledge that pays attention to process, incongruity, and the things we do not get to find out. The girl's clear knowledge is what directs her itinerary, while for me it is the itinerary, the process of looking for traces and directions on walls and sidewalks that directs my knowledge and shapes what I will know about these places.

I advance a few more metres along the street, and finally I see the Vicolo Castilia (see Figure 35). I spot a bicycle hanging from a street lamp on a house at mid-height, and then another one sticking out from the building wall close to the end of the street, and now it is easy to guess where my sought-after destination might be. I follow the wheels

Figure 35. Looking for the Stecca. From *top to bottom* and *left to right*: construction zone in the Isola; the "Cube"; wheels in mid-air; and posters on a wall. (May 2011).

in mid-air and come to the entrance of the Temporary Stecca. On the door, unfortunately closed, there is the program of events and many other posters. They refer to the Leoncavallo and other Social Centres and community associations. It is almost like a new city is opening its existence here. While I copy the events program into my diary, a middle-aged couple arrives, one of them is holding a flower vase. They, too, are looking for something. "It is closed," they exclaim. "It is beautiful, it is beautiful here," says the man to his companion and to me:

> They repair bicycles here. I have come here once, my daughter told me about it. She lives here, she came to live here recently.

And I see, literally right in front of me, the history of the space. I see the future in the shiny block that is supposed to be the new headquarters of the organization that once resided in an old disused factory here. I see on the land behind me the empty place where the Stecca had been. I see, where I am standing, its temporary home, embodying the fight for its existence. Even the posters, linking it to the broader contexts of occupations, of social and cultural struggles in Milan, have a long and complex story to tell. History seems inscribed in these two streets and their crossroads. As Harm observes, time is pasted onto urban territory, we can read it like a map.

Yet, reading these transformations, and the passage of time, like a clear map of how the city is growing seems to privilege one story over others. It echoes the voice of the flying girl who can see how things are going and why. The words of activists in the city who shared their knowledge with me – Alice and Giacomo, Flavia, Ottavia, Alberto, the Zenobia participants, and many others – suggest a different model of knowledge. I could describe it as a perspective from the interstices. Looking at the city from spaces deemed empty and useless, from the voids, activists from various organizations point out the discrepancies in the fabric of Milan. They ask: Why is it that in the city of cranes so many people cannot afford a home? Why is it that the city of fashionable appearances is creating so many zones of invisibilities? These interrogations are possible only because the interstices offer a partial vision. This is not the panoptic understanding of the one who flies where everything is visible. Rather, it is more akin to the layered speaking of the walls close to the Vicolo Castilia, with their thick skins of posters and inscriptions.

Walking along this wall, looking for Vicolo Castilia, I was hailed by their announcements: from events and public talks to important

municipal notices. There were instructions for applying for social housing, announcements of the elections, and a description of how to vote.
It is the citizen, the inhabitant, who is interpellated in this exchange.
Yet, this knowledge is literally on the walls, the buildings, the textures
of the city. It is steeped in everyday interactions, small exchanges like
the ones with the two women and the man with a camera, crossroads
and itineraries. Interrupted, partial, sometimes surprised or questioning, this mode of learning says something about the politics of space
that coherent plans and descriptions cannot.

If the poster, a long-standing, key element in Italian urban life and
politics, is here a useful metaphor for knowledge, political engagement,
or memory, it is because it is always ripped at the corner, ephemeral
because always pasted over, contested from the start, and only partially
legible. The poster recognizes that every clear plan might only be a temporary break from the constant battle of voices. When, a few days later,
I do find the Stecca open, I have a long conversation with Carlo, an
activist of the association, between mountains of bicycles and dozens of
people from the community busy repairing them. We are in the courtyard of the Temporary Stecca, onto which the library of the organization
Cantieri and a little café open. "Where are you going from here," I ask
him, "Aren't you moving to the new home?" "Oh *that* building," he tells
me. He recounts that with impending elections nobody knows what
is happening, who that building really is for, and whether they will
ever move there. Most importantly, he tells me that nobody even talks
about it. "The future is unknown," he announces, and everybody is just
waiting in anticipation of something to come. Nobody knows what it
is. "Everything is stopped," he says, at least until the election. People
might talk again about the future after that.

Herein lies the difficulty: what Carlo, and others can reply to the girl
is a silence, an unknown, a not-known yet. This is a hard response to
give to somebody so resolute and with such a clear vision. This discrepancy, however, calls attention to the very processes through which
spaces, landscapes, and social subjects are given meanings in contemporary Milan. It also calls for an ethnographic attention to the ways in
which stories emerge in particular places and moments that are always
connected to other tales, other encounters, and other locales. If we walk
around the fences of the Garibaldi Repubblica construction zone, we
might see that there is a history of space pasted onto the sidewalk. It
is only through tangled connections and itineraries, it is only through

searching for tracks that it becomes legible, apparent. And when it does, it opens more questions than answers: Why is the cube empty? If that is supposed to be the future, where is the story stuck? If that is supposed to be a public space, who will claim it as such and in what ways? What other understandings of piazzas and streets will shape Milan in the years to come? Who will be part of these "sand dunes," and for what purposes?

Glossary

Agora: The market square of the Greek city. More generally, this term refers to the idealized plaza where city dwellers can come together to voice their opinions, and to debate issues of importance to society. Active participation in the Greek agora was, however, reserved to men who were free (non-slaves) and "masters of households" (Habermas, 1989: 3).

Chronicles: Particularly situated recollections and analyses, aimed at reformulating the histories of places, people, and events in the city.

Dismissed areas (*aree dismesse*): Empty factory (or small manufacturing) buildings, vacant fields, and workers' facilities that have been "left over" from the process of deindustrialization. Train stations or depots that are no longer in use are included in this category as well. Some of these areas have been appropriated (legally and/or illegally) by organizations and groups ranging from political networks to artists and are used as meeting spaces, independent community centres, or worksites. Other dismissed areas are currently the sites of major redevelopment projects.

Extracomunitario: Literally, a person "outside of the community." Commonly used to refer to immigrants (and especially visible minority residents), this term designates people who migrate to Italy and who do not belong to the European Union.

Heteroglossia of vision: The way in which images and different modalities of looking, showing, and appearing co-inhabit urban locales. This term, borrowed from Bakhtin (1981: 263) emphasizes that public space is always already criss-crossed with visual acts.

Milanese: This term usually refers to people born in Milan, and, depending on the speaker, it can also include those who have lived in the city a significant number of years. Often "Milanese" is used in an exclusionary way, to mark degrees of entitlement to the city, and to discriminate against foreign immigrants and southern Italians. In this book, unless otherwise indicated, I use "Milanese" in an inclusive way, referring to all people who are currently residing in Milan.

Security Package (*pacchetto sicurezza*): A set of measures proposed in May 2008 by the Berlusconi government, which defines illegal immigration as a crime, making deportation easier and allowing longer periods of detention for immigrants without papers. In February 2009, other measures were added, including doctors being obliged to report illegal immigrants visiting hospitals, and the creation of vigilante patrols to monitor urban public spaces. Although the Security Package sparked widespread critiques, many of the proposed measures were approved by the Italian government in July 2009 (for more information, see Merlino, 2009).

Social Centres (*centri sociali*): Autonomous social and/or community centres that serve as venues for political, social, and countercultural grassroots activities. They are usually housed in "squatted" (illegally claimed and occupied) buildings. While each Centre forms a distinct entity, most of them are connected to each other and constitute a wider alliance and action network in Milan and in Italy.

***Struscio*:** Literally, the brushing of one's body among others, the term designates the promenade carried out in most Italian cities. It consists of leisurely strolling in the central streets to see people and show one's (good) appearance. While the *struscio* is much less prominent in contemporary Milan than it used to be, as an embodied and performative "visual intermingling" (Pinney, 2002: 364) in public space, it still informs visual practices in the city.

Precariousness (*precarietà*): The unregulated, temporary, and flexible working (and thus also living) conditions of many people in Italy, brought on in large part by neoliberalism and its deregulation of labour. More generally, this term describes social vulnerability and serves as an anchor for critiques of neoliberalism in Italy (Molé, 2010). Many Social Centres have been involved in actions and campaigns against precariousness.

Public space: In a general and broad sense, space that is accessible to all and that is publicly rather than privately owned, such as streets, piazzas, and parks. In this research, I refer to public spaces both as the specific material locales that my interlocutors referred to, and as spatial enactments that give a particular form and meaning to this concept, and thus serve as ethnographic entry points from where to trace local discourses, processes, and debates.

San Precario: The symbolic, "invented," patron saint of precarious (temporary, flexible, occasional, or sessional) workers. He is sometimes depicted
in the form of a statue in front of the Social Centre Casa Loca. As a circulating image, an installation, a carnivalesque figure, and a tool in rallies, San
Precario serves as a referent for critiques of neoliberalism and precarious
working conditions.

Serpica Naro: The "alter ego" of San Precario (and an anagram of the latter);
a designer invented by the association Chainworkers to draw attention
to the precarious labour conditions in the field of fashion. Serpica Naro
participated in the Fashion Week of spring 2005 mocking and criticizing the
fashion system, and the politics of appearance, gender, and labour relations
in contemporary Milan.

Notes

Introduction

1 With a "public" I refer here to a group of people who participate, in various ways, in public space. As Harms (2009) and Iveson (2007) point out, it is important to remember that there is not only one "public" that participates in streets and piazzas, but rather many, differently constituted and situated publics, whose interests, perspectives, and social identities might be quite dissimilar, or even in contraposition to each other.

2 According to Monteleone and Manzo (2010: 139), only 8% of its inhabitants are Chinese-Italians.

3 I thank one of my anonymous reviewers for this observation.

4 All translations of materials written or spoken in Italian – written publications, people's quotes and interviews, websites, flyers, and pamphlets – are mine.

5 The Canonica-Sarpi neighbourhood is an interesting example of these complexities. From the 1920s, Chinese small stores and artisans have used the workshops at the street level and the adjacent accommodations, perpetuating the historical mix between small manufacturing and residences characteristic of this area (Monteleone and Manzo, 2010). More recently, economic and trade changes, as well as a deregulation of municipal retail frameworks, have resulted in the displacement of the small businesses and the rise of the wholesale sector. At the same time, the area has been increasingly gentrified. Indeed, according to Monteleone and Manzo, it is the fascination for popular neighbourhoods such as these – with their turn-of-the-century apartment buildings and their mix of artisanal and living spaces – that has facilitated this process. As a result, it is now showing a peculiar vertical division. While more affluent, white, Italian-born residents live in the upper floors of the apartment

buildings, Chinese-Italian residents occupy the lower areas: it is a "spatial and symbolic separation between the neighbourhood 'high' and 'low'" (ibid.). The claims, by some Italian residents, that "Chinese people" are threatening the culture of local places and ruining familiar cityscapes effectively mask this layered inequality between inhabitants. And it ignores that half a century of living and working in the lower floors by the Chinese community had actually been more in tune with the history of the neighbourhood than the recent gentrification by the Italian middle and upper classes and the concurrent neoliberal deregulation of city areas (ibid.).

6 I owe this insight to an anonymous reviewer.

7 This anecdote was first published in Moretti (2008a).

8 The length of the interviews ranged from half an hour to two and a half hours. Most lasted between 45 and 60 minutes. In some interviews, I asked participants to draw a map of Milan, and of the public spaces they use; see, e.g., Tupeshka (2001). While in a few cases people drew detailed and elaborate maps of the city, in most cases mapmaking provided just a springboard for my interlocutors to reflect and comment on the city, its spaces, and their lives in it.

9 Some paragraphs from this section appeared in Moretti (2011).

10 I thank one of my anonymous reviewers for this expression and suggestion.

1. Orientations

1 The randomness of social interaction in urban space is particularly important, because it counters most people's propensity to be comfortable in one's familiar milieu. In this way, the public life of streets and piazzas of modern cities ideally forces people to relate to others on the basis of a common citizenship, participation in society, and equal human rights (cf. Caldeira, 2000: 303).

2 Significantly, where public space is curtailed, restricted, and privatized, this has negative consequences for civil society, justice, and participatory democracy (Falzon, 2004; Caldeira, 2000).

3 See, e.g., Crossley and Roberts (2004) and Guano (2002).

4 Pigg (1996: 164) discusses the points of view from where certain categories become visible, and the ways in which modernity changes meanings and roles depending on the social positions of those who use this term.

5 With "cosmopolitanism" I refer to an outward orientation that constructs what is seen as "global" and/or internationally trendy as a sign of distinction and privilege for its holder. As Schielke (2012), De Koning

(2009), and Tsing (2005) have discussed, cosmopolitanism is constructed differently in different settings, thus referring to various objects, lifestyles, and desires.

6 Some parts of this discussion on the *struscio* appear in Moretti (2008a).

7 These practices, moreover, not only can be seen as a burden for women, but also as a vehicle of sociality, cultural creativity, and subtle empowerment (Guano, 2007).

8 De Certeau distinguishes between *places* as hegemonically constructed locations, and *spaces* as inhabited, lived, and shaped by relations and interactions.

9 The sexualized component of the models' work also refers to the gender roles perpetuated by many fashion commercials, and to the very similar actions of sex workers – both Italian- and foreign-born – who also show their bodies to men and women travelling inside cars.

10 The pleasant surprise ("Modelle Lavavetri," 2007) of drivers watching the models relies on and strengthens the contrast between desirable and undesirable bodies in public space.

2. Milan

1 Some paragraphs in this chapter were first published in Moretti (2011).

2 The dance in the Galleria took place on 9 February 2009. The walk through the Giardini Pubblici (Public Gardens) and through Corso Buenos Aires is of 19 February 2009. I saw the violinist on the subway on 12 March 2009. My conversation in the University of Milan-Bicocca with a student researching deindustrialization is from 3 February 2009. My meeting a friend in Bicocca, walking with her to Casa Loca, and our conversation with members of Casa Loca took place on 7 April 2009. Eliza and I walked in Bovisa on 18 and 20 February 2009. I met the elderly woman with very blue eyes on 10 April 2009.

3 Foot (2001: 3), for example, calls Milan the "epicentre" of "all the crucial movements, booms, slumps, and moments in twentieth-century Italian history" (see also Muehlebach, 2012). For a description of the "miracle," and some major historical shifts in Milan, see Ginsborg (2003) and Foot (2001). See Petrillo (2004) and Parsi and Tacchi (2003) for a discussion of the role of the factories in Milan.

4 Futurism is a cultural movement that developed in Italy, and in Milan particularly, at the beginning of the twentieth century. Guided by Marinetti as one of its most distinguished interpreters, futurists celebrated the upcoming modernity as an age of machines, speed, strength, and virile

energy, which would lead Italian society into a new era. The events of 2009 marked the 100 years that have passed since the publication of Marinetti's "Futurist Manifesto."

5 Together with its greater metropolitan region, Milan counts about four million inhabitants (Aalbers, 2007). The architect and expert in urban planning Paola Rottola, one of the people I met in Milan, estimates that the number of people who inhabit Milan during the day is more than three million.

6 The Lombardy Region, moreover, is a major player in the European and the Italian economy, with "40 percent of the country's businesses, half of its jobs in the information sector, and 30 percent of all research and development" (Muehlebach, 2012: 14).

7 The lack of affordable housing, be it rented or owned, has in part been a result of recent trends in the real estate market and the limited availability of rental accommodations. As Aalbers (2007) discusses, the recent dynamism of the Milanese real estate market has generally rendered housing less affordable for Milanese residents, resulting in individuals and families paying more for accommodations, and incurring greater risks than before as they are forced to take greater loans to pay for it. As far as rental accommodations are concerned, the choice has always been very limited. Milan has a very high percentage of home ownership, also due to laws that make owning a house very advantageous and at the same time pose very strict conditions on renting space to tenants. Unfortunately, several reforms that were implemented in the past 40 years as a way to ensure affordable rents have had almost opposite results, such as increased evictions of tenants (ibid.). Intergenerational transfer of real estate plays an important role in this equation. Young families are often, if possible, helped by their parents in acquiring a home. Coupled with the general lack of rental housing in Milan, this makes the disparity between more and less advantaged sectors of the population even more pronounced. In the words of Aalbers (ibid.: 184), in this context "housing also deepens and structures existing social and economic inequalities."

8 Caritas is a Catholic organization, funded by the Catholic Church, promoting social development and offering assistance to marginalized people. It publishes every year a comprehensive statistical dossier on immigration in Italy.

9 The shift is from an economy centred on manufacturing to one based on services.

10 Zajczyk and Cavalca (2006) describe the feminization of poverty in the city, with single mothers and elderly women living alone being particularly at

risk. The latter group, for example, makes up about a third of the people in poverty, although they only count for 23.5% of the population (Benassi, 2005) – a fact due both to women's longevity and their lesser participation in the workforce during their lifetime.

11 See, e.g., Stacul (2007) and Comaroff and Comaroff (2009).

12 They could not take into account, for example, how the perceived general sense of immobility in Milan is actually "the result of a molecular frenzy" (Boeri, 2007: 9), and continuous diffuse movements. And Doninelli (2010: 169) writes that the difficulty of representing and explaining Milan comes from the coexistence of "multiple destinations, intersecting flows" and directions within a city that nonetheless strives to live in a single shared dimension.

3. The Agora of the City

1 As a virtual piazza, VivereMilano assembles people who live all over the city, and whose busy schedules might not otherwise allow them to talk and listen to each other. Once people become readers/bloggers, they can respond to others' invitations to go to particular meetings and events and/or decide to join or form an interest group.

2 This comment refers to three cases, in the preceding few months, of elderly residents being discovered dead in their apartments because of the social isolation in which they lived.

3 As E. Guano (personal communication, 28 Mar. 2008) points out, this "mandatory leisure" has led to the "cementification" of many places and booming tourism in the nearby regions. In an additional ironic turn, some of the 30-year-olds might be especially eager to leave the city on the weekends if they are living with their parents, and/or might be able to do so precisely because they can use their parents' recreational houses at the seaside or in the mountains.

4 In the following sections, I speak about a-politics as the avoidance of political engagement, actions, and discussions, with politics understood as a formal, institutionalized realm, shaped by the electoral process and expressed by political parties. I refer to anti-politics as an active refusal and distancing from state political institutions and processes.

5 In cities with more than 15,000 inhabitants, if none of the candidates wins an absolute majority of votes, another round of voting has to take place two weeks later, with voters choosing between the two candidates who obtained the most votes in the first round.

4. Spatial Politics

1 See Tonnelat (2008), Rademacher (2008), and Dines (2002) for examples of disorderly spaces and people in Paris, Kathmandu, and Naples, respectively.
2 Estimates of the numbers of Social Centres are often tentative, as many of them have short lives. New Centres are opened all the time, as old ones close, move/are dislocated, and/or resurface in different forms.
3 See Vecchi (1994) and Quagliata (1994) on Social Centres escaping definitions and exceeding sociological theorizing.
4 Resistance against different form of fascism has always been an integral part of the existence of Social Centres. This fundamentally political and historical conflict has often been accompanied by police repression and by violent confrontations between right-wing street groups and the Social Centres. Moroni (1989), for example, talks about a veritable war within the city, and Alice recalls, "Here the killings are multiplying. The ones of Genova, and these, and it is not a small thing, starting from Fausto and Iaio, […] At the Leoncavallo we built a tree with the losses of the Leoncavallo [youth], and it was full, a tree full of photographs, full eh? Of the youth of the Social Centres that have been killed. That have been killed, not that have died" (18 Jan. 2005).
5 See also Quagliata (1994) for a history of the destruction and reconstruction of the Leoncavallo.
6 Vecchi (1994: 5), for xample, calls them "fragments of an alternative public sphere," because they are both experimental places for creating social relationships based on equality and diversity, as well as spaces from which to debate urban spatial politics and governance.
7 This neighbourhood, which also includes the remaining Navigli canals, has always been "a mixed social composition of handicrafts, diffused factories, popular classes and extralegal" people and activities (Moroni, 1996: 167).
8 This is also the time of the "historic deal" between the left and the Christian Democratic Party, which was greatly disadvantageous for the institutional left (see Ginsborg, 2003).
9 As an example, Giacomo described a community initiative that he hoped would happen in the near future: a participatory "active mapping" project of Milan, which would be available online and would include the "social history of political struggles," the extent and effects of urban redevelopment, and alternative proposals (11 Mar. 2005).

5. Creating Spaces, Constructing Selves

1 Interestingly, De Giorgi (2010) points out that migrants work at the bottom of both the legal and the illegal economies.

2 In examining the shifts in the idea of European citizenship, Hansen (2000: 153–3) points out how the latter is increasingly defined along the lines of Christianity, as well as the legacy of historical moments such as the Enlightment, the Renaissance, and the "worldwide dissemination of European currents during the nineteenth century."

3 Calavita (quoted in De Giorgi, 2010: 159) describes this very poignantly: "Immigrants are useful as 'Others' who are willing to work, or are compelled to work, under conditions and for wages that locals now largely shun. The advantage of immigrants for these economies resides precisely in their Otherness. At the same time, that Otherness is the pivot on which backlashes against immigrants turn. For, if marginalized immigrant workers are useful in part because they are marked by illegality, poverty, and exclusion, this very marking, this highlighting of their difference, contributes to their distinction as a suspect population."

4 De Giorgi (2010), for example, points out that often the mere presence of immigrants in public spaces is interpreted as a sign of the decline in security and in the quality of life in a neighbourhood.

5 Church spaces that work as community centres are widely used in the city, and are known as *oratori*. They usually consist of a section of the church building and its courtyard, and serve as places of religious education, recreation facilities, headquarters for social services, cinema halls, and more, and range from conservative to progressive. Adults and children residing in the area of the parish can use these spaces and facilities to play, to socialize, and to organize specific activities.

6 Children born in Italy to "foreign" parents (i.e., residents who are not Italian citizens) have to wait until their eighteenth birthday to gain Italian citizenship.

7 To say it simply, Christian society would mirror itself and be represented in inclusive and egalitarian public space. The latter is then a place of transformation, because it can facilitate the realization of an ideal community in the here and now of our daily life. This was echoed also by the other young priests I met during my research, and reflected in the fact that they were all keenly interested in public space.

8 It is in the Brera area, close to the centre (see Figure 1 in the introduction). This area, traditionally known for its many cafés and the artists who lived

and worked there, is now a very expensive, elegant zone, with many clothing and fashion accessories boutiques.

6. Entangled (In)visibilities

1 Of the four organizations, Terre di Mezzo is the one I know best. In 2004–05 I became a kind of intern for this organization, and when I was not doing fieldwork in the streets, I was doing research from their offices. I participated in Terre di Mezzo events, and held long conversations as well as more formal interviews with some of its members. I met Naga through Terre di Mezzo and encountered its volunteers at the homeless initiatives of the latter. While I did not participate in the everyday operations of this group, I interviewed two of its members, and used its publications and reports. I visited the Stecca degli Artigiani several times, both in 2005 and in 2011, and held informal conversations as well as one formal interview with some of their members. Regarding "Building Zenobia," I did not participate in the actual workshop, but I attended the public presentation of the projects and interviewed one of its participants.
2 Consider the example of people migrating without papers across the Mediterranean. Their boats are at the same time metaphorically and literally not seen by ships that could intercept them, but also rescue them in an emergency, and this has had the most tragic consequences for hundreds of people.
3 Unfortunately, I could not attend this event. My information on it is based on the written description from the flyers, and the conversations with some of the activists at the Stecca degli Artigiani prior to the event.
4 In the words of Zenobia, "This kind of reading, we did it starting from the fact that we need another city, we need to construct an alternative, an alterity within this city, [...] so that inside this city mechanisms could proliferate and be born, that would allow the reoccupation of these empty areas. But empty not in the sense that there is nothing inside, but empty of meaning [...] What we want to give to the city is a deep meaning, a sense that we can participate, we can regenerate the territory" (Zenobia presentation, 9 Mar. 2005).
5 What I find interesting in this context is how the Internet allows for different modalities of seeing and showing, and thus for different representations of spaces and of the social relations that shape and are shaped by them. San Precario's virtual aspect shapes his place in streets, piazzas, supermarkets, and Social Centres, and his sudden surprising appearances in public spaces also serve to direct people to information and debates that are not immediately noticeable in piazzas and streets.

7. **Walking with Women: Vision and Gender in the City**

1 A version of this chapter was first published as Cristina Moretti (2008a), "A Walk with Two Women: Gender, Vision and Belonging in Milan, Italy," in Judith N. DeSena (ed.), *Gender in an Urban World* (Research in Urban Sociology, vol. 9), Emerald Group Publishing Ltd, 53–75. Reprinted with permission.

2 Several of my interlocutors, and these two women in particular, told me the story of their lives by describing and explaining some of the most important events they experienced, and by representing it as a meaningful chronological narration. The dialogues and walk we engaged in together became thus an opportunity for reflecting on their lives and the way in which they unfolded in the context of their families, friends, and the wider society.

3 As Granata et al. (2003) discuss, these churches have been lent to the Filipino-Italian community by the local church administration. Indeed, in recent years, priests who have immigrated from the Philippines, Latin America, and other countries have been working with immigrant communities all over Milan. According to Granata et al. (2003), the use of Catholic churches by Filipino-Italian people results in a certain autonomy of the community, which can organize its own festivities, events, and religious practices, while at the same time constituting a material link with Italian-born employers. The latter, in fact, can contact the churches or parishes to find a person to hire. This "permits members of the community to maintain a privileged position in the caretaking and service market" (147). At the same time, however, this creates a particular mix of visibility and invisibility in the city. While the Filipino-Italian community can congregate in these churches, it can never really claim them as their own, as they remain under local Milanese authority (ibid.).

4 Bourdieu (1984: 232) writes in this regard, "Choosing according to one's tastes is a matter of identifying goods that are […] attuned to one's position."

5 Milanese is currently spoken mostly by elderly people, although some middle-aged persons may also use it. It is very rare to find younger people who can understand it or speak it. An elderly woman I interviewed, who had spoken it at home with her parents and still used it exclusively with her husband, described it as the language of the working class, and remembered that all the upper-class families with whom her parents worked spoke only Italian. Today, however, Milanese is much less associated with class than with Milanese identity.

6 Women often shop for other members of their families, too, or accompany them if they need to buy clothing items. In addition, women do most

if not all of the washing, ironing, mending, and dry cleaning for their households. Women's work also includes the skilful mastering of sales, of alternative shopping circuits such as open air markets and factory outlets, and the use of informal trading networks.

7 According to Muehlebach (2012: 201), "Italy, together with Spain, is the European country with the highest share of immigrant women caring for the elderly."

8 As Zukin (1995) discusses with regard to New York City, gentrification is often linked with the creation of visual landscapes that appeal to middle- and upper-class viewers.

9 Indeed, my own ability to meet with these women in public spaces, to be friends with both of them, was exactly dependent on my ambivalent and contradictory "Milanese-ness" and on my status as in/outsider to both women, in different ways. As a woman born and raised in Milan by Milanese parents, I was for Francesca "Milanese" enough that she could talk about it with me. At the same time, my research and my living away from the city made me into a potential student, somebody to whom to show what "Milanese-ness" is. For Maria Anacleta, my own status as immigrant and my having a divided family myself made me into an ally and a friend: the fact that I am not simply nor really "Milanese" made it easier for her to talk to me; at the same time, as an Italian-born friend, I also represented one more connection to the city for her.

8. Places and Stages: Vision and Performance in Public Space

1 This chapter was first published as Cristina Moretti (2008b), "Places and Stages: Performing and Narrating the City in Milan, Italy," *Liminalities: A Journal of Performance Studies*, Special Issue on the City 4 (1): 1–46.

2 While the stage directions are in this sense fictional, we cannot say that the pictures are necessarily more "real," transparent, or authentic representations. As Pink (2007) and others remind us, photographs are selected, framed, edited, etc. Indeed, the double representation of the setting is meant to remind the reader about the role of narration – both Mohamed Ba's during the tour, and mine in this chapter – to construct frameworks and contexts for cultural commentary.

3 With "affect" here I do not intend an esoteric feeling contained in people's heads and minds. Emotions are important social constructions (Rudnyckyj, 2011; Navaro-Yashin, 2009; Abu-Lughod, 2002; Thrift, 2003; Mankekar, 1999). As they are responses to the complex and unequal social relations we are embedded in, they are never innocent or simply "beyond"

culture. In turn, they shape the political, cultural, and social choices we make, and the ways in which we approach and understand the world.

4 The use of different subjects for a recent Senegalese immigrant such as "(s)he," "one," "the immigrant," "our friend," "we"/"us," and "I" during the narration is particularly interesting, as it often shows a switch between different "voices." In this particular scene, for example, Mohamed's switch from "the immigrants" to "the immigrant" to "(s)he," "one," and "our poor immigrant friend" also signals a movement from a more detached to a more intimate perspective. What I translate as "(s)he" and "one" are important media of these shifts, because they do not indicate who the subject really is, but leave it open to the listener's interpretation. They often seem to beg the questions: how close is this "friend"? Is "one" just any-one or one-self? This is especially so with "(s)he." In Italian, subjects are not always required: although it is always possible, and sometimes necessary, to specify it, the subject is included in the conjugated verb form. For example, *arriva* means (he? she?) arrives. Many of Mohamed's sentences have such "absent"/unspecified subjects: I translate them as "(s)he." (I translate them as (s)he even when the use of pronouns earlier in a particular passage suggests that it is a male character.)

5 I constructed a chorus out of these and other lines because of the ways in which these sentences were spoken. In Mohamed's narration, they took on the roles of admonitions, advice, and insights for the audience.

6 The ambiguity of the subject in this passage is another example of the alternating and interweaving of speaking positions: While "us" refers to the people who have lived in Milan for a long time, thus distancing the speaker from the "(s)he" "who arrives," the "our" in the line "they are eaten in our side [of the world]" denotes a speaker who is very close and familiar to the "(s)he" who is told: "No, no, no, one does not touch them, one does not eat them." So here again: who is "(s)he"? And who is "one"?

7 Here is an example of several itineraries/paths intersecting: the course of the canals, the exploring paths of a newcomer through the city, international migrations, the tour of Mohamed and me through the city.

8 This is another interesting example for the coexistence of several "voices." It could be rewritten like this:
NARRATOR: Following the course of the water, (s)he realizes that it becomes a belt, a belt that suddenly ends behind here […]
THE FRIEND: but how come [we] ended [here], in a street which on the other side is called Via Laghetto [Little Lake Street]? […] It is strange, because here there isn't any water!

THE VOICE OF HISTORY: But yes, there was water: water underneath [...]
NARRATOR: So (s)he continues his (her) tour and cannot see another way
    than the Duomo.

9 The water, reaching Laghetto Street (Via Laghetto) and the Little Lake
   of Santo Stefano (Laghetto di Santo Stefano) carried boats with building
   materials for the construction of the Duomo. According to the association
   Friends of the Navigli (www.amicideinavigli.it), navigation to carry
   marble for the Duomo started in 1387. Leonardo da Vinci is also associated
   with the Navigli waterways, because he designed their lift locks (the
   Chiuse di Leonardo; see Figure 30 in scene II). The part of the Navigli
   canals closer to the city centre, called *cerchia interna*, was entirely covered
   in the 1930s. Some of the canals that were further from the centre (such as
   parts of the Naviglio Grande and Naviglio Pavese) are still visible today.
   The construction of the Duomo lasted from 1386 to the nineteenth century,
   earning the cathedral the reputation of a never-ending work in progress.
10 In the novel *Marcovaldo*, Calvino (1963) writes about a city dweller trying
   to catch pigeons to prepare a tasty meal.
11 "O mia bela madunina" is a very popular old song often used as a symbol
   of Milan, together with the golden statue of Mary to which it refers.
12 See Fo (1974).
13 In our actual tour, at this point Mohamed walked away from the
   courtyard, returned to Dante Street (where we were in the previous scene),
   and started again walking away from the Duomo. He headed towards the
   Sforza Castle, which is about a ten-minute walk away.
14 The three examples Mohamed talks about here are considered traditional
   music forms from southern Italy. The *pizzica* and the *tarantella* are linked to
   tarantismo, a system of trance-inducing curing rituals for people who have
   been bitten by Tarantula spiders. I find very interesting that the examples
   Mohamed uses for cultural identity and difference are musical, something
   that is very processual, and performative.
15 In our actual itinerary, here Mohamed started to go towards nearby
   Cadorna Station, where the rest of the narration took place.
16 In this part of his narration, Mohamed also describes how a vendor goes
   from one train station (Cadorna) to another (Central Station) to sell his or
   her merchandise, following the commuters into and out of Milan.
17 See Thrift (2003: 2021) for a discussion on the methodological and
   theoretical importance of the "depth of the now."
18 Taylor (1997: 30, quoting Derrida) writes: "Derrida [...] highlights the
   reiterative nature of haunting, for phantoms always represent a repetition:
   'A specter is always a *revenant*. One cannot control its comings and goings
   because it *begins by coming back*'" (emphasis in Derrida's original).

Conclusion: Into the Future

1   It is interesting that in the English version of the video (available at the
    Internet site of Porta Nuova www.porta-nuova.com), the girl's narration
    is the only translated part. Her description constitutes the beginning
    and ending sections of the film, with seven excerpts of interviews with
    architects and developers in the middle. Three of these are in English,
    and four in Italian. Interestingly, the English-speaking interviews are not
    subtitled in the Italian version. Considering the pervasive presence in the
    video of high-tech landscape simulations of places and buildings that do
    not exist yet, it is hard to imagine that subtitling would have been too
    expensive or complicated to add. Similarly, in the English version, none
    of the Italian interviews are translated or subtitled, leaving the audience
    to hear Italian explanations for long minutes. The impression, then, is
    that the filmmaker simply did not bother to render understandable the
    promoters' explanations and messages. In contrast, the whole text of the
    girl has been translated. This particular linguistic choice strengthens the
    sense that the girl is the key element of the video, because of the particular
    sensibility towards the future that she embodies.

Bibliography

Aalbers, Manuel B. 2007. "Geographies of Housing Finance: The Mortgage Market in Milan, Italy." *Growth and Change* 38 (2): 174–99. http://dx.doi.org/10.1111/j.1468-2257.2007.00363.x.

Abu-Lughod, Lila. 2002. "Egyptian Melodrama." In *Media Worlds: Anthropology on New Terrain*, ed. Faye D. Ginsburg, Lila Abu-Lughod, and Brian Larkin, 115–33. Berkeley: University of California Press.

Acanfora, Massimo. 2004. "La casa dentro." *Terre di Mezzo*: 4–5.

Acanfora, Massimo. 2003. "Gente in bilico." *Terre di Mezzo*: 15.

Amin, Ash. 2004. "Multiethnicity and the Idea of Europe." *Theory, Culture & Society* 21 (2): 1–24. http://dx.doi.org/10.1177/0263276404042132.

Amireaux, Valerie. 2006. "Speaking as a Muslim: Avoiding Religion in French Public Space." In *Politics of Visibility: Young Muslims in European Public Spaces*, ed. Gerdien Jonker and Valerie Amireaux, 21–52. Bielefeld: Transcript Verlag.

Anderlini, Fausto, and Matteo Bolocan Goldstein. 2011. "Milano, Italia: Segnali dal Nord." *Il Mulino* 4: 578–89.

Anderson, Bridget. 1999. "Overseas Domestic Workers in the European Union." In *Gender, Migration and Domestic Service*, ed. Janet Henshall Momsen, 117–33. London: Routledge. http://dx.doi.org/10.4324/9780203452509_chapter_7.

Andreotti, Alberta. 2006. "Coping Strategies in a Wealthy City of Northern Italy." *International Journal of Urban and Regional Research* 30 (2): 328–45. http://dx.doi.org/10.1111/j.1468-2427.2006.00669.x.

Ansa News. 2006, 11 Mar. "Centri sociali: Momenti di forte tensione a Milano." www.ansa.it

Appadurai, Arjun. 1996. *Modernity at Large: Cultural Dimensions of Globalization*. Minneapolis: University of Minnesota Press.

Armentano, Angelo, and Valeria Lupatini. 2007. "Trenta grandi trasformazioni" In *Milano incompiuta: Interpretazioni urbanistiche del mutamento*, ed. Matteo Bolocan Goldstein and Bertrando Bonfantini, 65–96. Milan: Quaderni del Dipartimento di Architettura e Pianificazione.

Bakhtin, M.M. 1981. *The Dialogic Imagination: Four Essays*. Ed. Michael Holquist. Trans. Caryl Emerson and Michael Holquist. Austin: University of Texas Press.

Balducci, Alessandro. 2007. "Quale rinascimento per Milano?" In *Milano incompiuta: Interpretazioni urbanistiche del mutamento*, ed. Matteo Bolocan Goldstein and Bertrando Bonfantini, 7–8. Milan: Quaderni del Dipartimento di Architettura e Pianificazione.

Banks, Marcus, and Howard Morphy, eds. 1997. *Rethinking Visual Anthropology*. New Haven: Yale University Press.

Bauman, Richard, and Charles Briggs. 1990. "Poetics and Performances as Critical Perspectives on Language and Social Life." *Annual Review of Anthropology* 19 (1): 59–88. http://dx.doi.org/10.1146/annurev.an.19.100190.000423.

Bayat, Asef. 2012. "Politics in the City-Inside-Out." *City & Society* 24 (2): 110–28. http://dx.doi.org/10.1111/j.1548-744X.2012.01071.x.

Benassi, David. 2005. "La povertà in un contesto ricco: I milanesi poveri." In *La Povertà come condizione e come percezione: Una survey a Milano*, ed. David Benassi, 15–36. Milan: FrancoAngeli.

Benjamin, Walter. 1969. *Illuminations*. New York: Schocken.

Berdahl, Daphne. 1999. "'(N)Ostalgie' for the Present: Memory, Longing, and East German Things." *Ethnos: Journal of Anthropology* 64 (2): 192–211. http://dx.doi.org/10.1080/00141844.1999.9981598.

Blomley, Nicholas. 1998. "Landscapes of Property." *Law & Society Review* 32 (3): 567–612. http://dx.doi.org/10.2307/827757.

Blomley, Nicholas. 1996. "I'd Like to Dress Her All Over: Masculinity, Power, and Retail Space." In *Retailing, Consumption, and Capital: Towards the New Retail Geography*, ed. Neil Wrigley and Michelle Lowe, 238–56. Harlow: Longman.

Boeri, Stefano. 2007. "Caleidoscopio Milano." In *Milano: Cronache dell'abitare*, Multiplicity.lab, 9–17. Milan: Mondadori.

Bondi, Liz, and Mona Domosh. 1998. "On the Contours of Public Space. A Tale of Three Women." *Antipode* 30 (3): 270–89. http://dx.doi.org/10.1111/1467-8330.00078.

Bonomi, Aldo. 2008. *Milano ai tempi delle moltitudini*. Milan: Mondadori.

Boschetti, Andrea. 2010. "Metrogramma B&F." *Dedalo* 20: 26–7.

Bourdieu, Pierre. 1984. *Distinction: A Social Critique of the Judgement of Taste*. Cambridge: Harvard University Press.

Braghiroli, Stefano. 2011. "The Italian Local Elections of 2011: Four Ingredients for a Political Defeat." *Bulletin of Italian Politics* 3 (1): 137–57.

Brenner, Neil. 2004. *New State Spaces: Urban Governance and the Rescaling of Statehood*. Oxford: Oxford University Press. http://dx.doi.org/10.1093/acpr of:oso/9780199270057.001.0001.

Brenner, Neil. 1998. "Global Cities, Glocal States: Global City Formation and State Territorial Restructuring in Contemporary Europe." *Review of International Political Economy* 5 (1): 1–37. http://dx.doi.org/10.1080/096922998347633.

Breveglieri, Lorenzo, Daniele Cologna, and Giovanna Silva. 1999. "Realtá, limiti e potenzialitá dell'integrazione culturale Afro-Milanese." In *Africa a Milano: Famiglie, ambienti e lavori delle popolazioni africane a Milano*, ed. Daniele Cologna, Lorenzo Breveglieri, Elena Granata, and Christian Novak, 49–72. Milan: Abitare Segesta.

Bricocoli, Massimo, and Paola Savoldi. 2010. "Urbanistica e politiche alla prova dei luoghi." In *Milano Downtown: Azione pubblica e luoghi dell'abitare*, ed. Massimo Brococoli and Paola Savoldi, 191–231. Milano: Et al. Edizioni.

Bridgman, Rae. 2006. *StreetCities: Rehousing the Homeless*. Toronto: Broadview.

Burigana, Alessandra. 2009. "La Grande Agorá dei Milanesi." *Casamica* (1/2): 25–7.

Burk, Adrienne. 2010. *Speaking for a Long Time: Public Space and Social Memory in Vancouver*. Vancouver: UBC Press.

Caldeira, Teresa Pires do Rio. 2000. *City of Walls: Crime, Segregation, and Citizenship in São Paulo*. Berkeley: University of California Press.

Calhoun, Craig. 1992. "Introduction: Habermas and the Public Sphere." In *Habermas and the Public Sphere*, ed. Craig Calhoun, 1–48. Cambridge: MIT Press.

Calvino, Italo. 1981 [1979]. *If on a Winter's Night a Traveler*. New York: Harcourt Brace Jovanovich.

Calvino, Italo. 1963. *Marcovaldo, ovvero le stagioni in città*. Turin: Einaudi.

Caritas/Migrantes. 2008. *Dossier Statistico Immigrazione*. Rome: Caritas.

Caritas/Migrantes. 2005. *Dossier Statistico Immigrazione*. Rome: Caritas.

Caronia, Antonio. 2005, 4 Mar. *Progetto Zenobia: Il diritto all'abitare fra locale e globale*. http://www.socialpress.it.

Carter, Donald Martin. 1997. *States of Grace: Senegalese in Italy and the New European Immigration*. Minneapolis: University of Minnesota Press.

Casa Loca Website. http://www.casaloca.it.

Casamica (weekly magazine of *Corriere della Sera*). 2009. Milan: Domani.

Castañeda, Quetzil. 2006. "The Invisible Theatre of Ethnography: Performative Principles of Fieldwork." *Anthropological Quarterly* 79 (1): 75–104. http://dx.doi.org/10.1353/anq.2006.0004.

Castellaneta, Carlo. 1997. *Nostalgia di Milano*. Milan: Mondadori.

Chainworkers Website. www.chainworkers.it; accessed Sept. 2007.

Checker, Melissa. 2011. "Wiped Out by the 'Greenwave': Environmental Gentrification and the Paradoxical Politics of Urban Sustainability." *City & Society* 23 (2): 210–29. http://dx.doi.org/10.1111/j.1548-744X.2011.01063.x.

Chiaramonte, Alessandro, and Roberto D'Alimonte. 2012. "The Twilight of the Berlusconi Era: Local Elections and National Referendums in Italy, May and June 2011." *South European Society & Politics* 17 (2): 261–79. http://dx.doi.org/10.1080/13608746.2012.701793.

Chiari, Matteo. 2004. "Stranieri on the Road." *Terre di Mezzo*: 16–17.

Ciorra, Pippo. 2003. "Adriaticitta." In *Italian Cityscapes: Culture and Urban Change in Contemporary Italy*, ed. Robert Lumley and John Foot, 199–203. Exeter: University of Exeter Press.

Cole, Jeffrey, and Sally Booth. 2007. *Dirty Work: Immigrants in Domestic Service, Agriculture, and Prostitution in Sicily*. Plymouth: Lexington Books.

Cologna, Daniele. 2003a. "Profilo sociografico delle principali popolazioni asiatiche a Milano." In *Asia a Milano: Famiglie, ambienti e lavori delle popolazioni asiatiche a Milano*, ed. Daniele Cologna, 24–60. Milan: Abitare Segesta.

Cologna, Daniele, ed. 2003b. *Asia a Milano: Famiglie, ambienti e lavori delle popolazioni asiatiche a Milano*. Milan: Abitare Segesta.

Cologna, Daniele, Lorenzo Breveglieri, Elena Granata, and Christian Novak, eds. 1999. *Africa a Milano: Famiglie, ambienti e lavori delle popolazioni africane a Milano*. Milan: Abitare Segesta.

Comaroff, John, and Jean Comaroff. 2009. *Ethnicity, Inc.* Chicago: University of Chicago Press. http://dx.doi.org/10.7208/chicago/9780226114736.001.0001.

Comaroff, Jean, and John Comaroff. 2003. "Ethnography on an Awkward Scale: Postcolonial Anthropology and the Violence of Abstraction." *Ethnography* 4 (2): 147–79. http://dx.doi.org/10.1177/14661381030042001.

Committee for Letizia Moratti Mayor of Milan. 2011. "I cento progetti realizzati." Booklet produced by the committee.

Comune di Milano. 2004 http://www.comune.milano.it/statistica/index.html.

Crane, Diana, and Laura Bovone. 2006. "Approaches to Material Culture: The Sociology of Fashion and Clothing." *Poetics* 34 (6): 319–33. http://dx.doi.org/10.1016/j.poetic.2006.10.002.

Crossley, Nick, and John M. Roberts, eds. 2004. *After Habermas: New Perspectives on the Public Sphere*. Oxford: Blackwell.

Dal Lago, Alessandro. 2009. *Non-Persons: The Exclusion of Migrants in a Global Society*. Trans. Marie Orton. Milan: IPOC.

Dal Lago, Alessandro. 1999. *Non-Persone: L'Esclusione dei migranti in una società globale*. Milan: Interzone.

de Certeau, Michel. 1984. *The Practice of Everyday Life*. Berkeley: University of California Press.

De Giorgi, Alessandro. 2010. "Immigration Control, Post-Fordism, and Less Eligibility: A Materialist Critique of the Criminalization of Immigration across Europe." *Punishment and Society* 12 (2): 147–67. http://dx.doi.org/10.1177/1462474509357378.

De Koning, Anouk. 2009. *Global Dreams: Class, Gender, and Public Space in Cosmopolitan Cairo*. Cairo: American University in Cairo Press.

De Lucchi, Michele, and Andrea Villani. 2004. "Aprire e chiudere: Lo scenario per le scelte." In *Sulla città, oggi: La nuova piazza*, ed. Giancarlo Mazzocchi and Andrea Villani, 73–94. Milan: FrancoAngeli.

Del Negro, Giovanna. 2004. *The Passeggiata and Popular Culture in an Italian Town: Folklore and the Performance of Modernity*. Montreal: McGill-Queen's University Press.

Derrida, Jacques. 1976. *Of Grammatology*. Trans. Gayatri Chakravorty Spivak. Baltimore: Johns Hopkins University Press.

Devoti di San Precario. 2004, 21 Sept. "Milano ModaDonna 2004? No Peace, No Party!" http://www.socialpress.it.

Di Leonardo, Micaela. 1987. "The Female World of Cards and Holidays: Women, Families, and the Work of Kinship." *Signs* 12 (3): 440–53.

Dines, Nicholas. 2002. "Urban Renewal, Immigration, and Contested Claims to Public Space: The Case of Piazza Garibaldi in Naples." *GeoJournal* 58 (2/3): 177–88. http://dx.doi.org/10.1023/B:GEJO.0000010837.87618.59.

Dines, Nicholas. 1999. "Centri sociali: Occupazioni autogestite a Napoli negli anni novanta." *Quaderni di Sociologia* 43 (21): 90–111.

Dolan, Jill. 2005. *Utopia in Performance: Finding Hope at the Theater*. Ann Arbor: University of Michigan Press.

Domosh, Mona. 1996. "The Feminized Retail Landscape: Gender, Ideology, and Consumer Culture in Nineteenth-Century New York City." In *Retailing, Consumption, and Capital: Towards the New Retail Geography*, ed. Neil Wrigley and Michelle Lowe, 257–70. Harlow: Longman.

Doninelli, Luca. 2010. "L'alba del degrado." In *Milano é una cozza*, ed. Luca Doninelli, 161–71. Milan: Guerini e Associati.

Drieskens, Barbara, Franck Mermier, and Heiko Wimmen, eds. 2007. *Cities of the South: Citizenship and Exclusion in the 21st Century*. London: SAQI.

Entwistle, Joanne. 2000. *The Fashioned Body: Fashion, Dress, and Modern Social Theory*. Cambridge: Polity Press.

Fabian, Johannes. 1990. *Power and Performance: Ethnographic Explorations through Proverbial Wisdom and Theater in Shaba, Zaire*. Madison: University of Wisconsin Press.

Falzon, M. 2004. "Paragons of Lifestyle: Gated Communities and the Politics of Space in Bombay." *City & Society* 16 (2): 145–67. http://dx.doi.org/10.1525/city.2004.16.2.145.

Fausto e Iaio Website. http://www.faustoeiaio.org.

Fife, Wayne. 2004. "Semantic Slippage as a New Aspect of Authenticity: Viking Tourism on the Northern Peninsula of Newfoundland." *Journal of Folklore Research* 41 (1): 61–84. http://dx.doi.org/10.2979/JFR.2004.41.1.61.

Fikes, Kesha. 2009. *Managing African Portugal: The Citizen-Migrant Distinction.* Durham: Duke University Press. http://dx.doi.org/10.1215/9780822390985.

Fleetwood, Nicole. 2004. "'Busing It' in the City: Black Youth, Performance, and Public Transit." *Drama Review* 48 (2): 33–48. http://dx.doi.org/10.1162/105420404323063382.

Fo, Dario. 1974. *Morte accidentale di un anarchico.* Turin: Einaudi.

Foot, John. 2001. *Milan since the Miracle: City, Culture, and Identity.* Oxford: Berg.

Foschini, Paolo. 2005, 7 May. "La Prefettura: Venti agenti al Montestella e al Cassinis." *Corriere della Sera*: 51.

Fracca, Cesare. 2006, 22 May. VivereMilano Website. http://www.vivere.milano.it; accessed June 2006.

Freeman, Richard. 2001. "The City as Mise-èn-Scene: A Visual Exploration of the Culture of Politics in Buenos Aires." *Visual Anthropology Review* 17 (1): 36–59. http://dx.doi.org/10.1525/var.2001.17.1.36.

Frisina, Annalisa. 2006. "The Invention of Citizenship among Young Muslims in Italy." In *Politics of Visibility: Young Muslims in European Public Spaces*, ed. Gerdien Jonker and Valerie Amireaux, 79–101. Bielefeld: Transcript Verlag.

Gagliardi, Andrea. 2007, 10 July. "Parchi blindati a Milano. Gli immigrati: No ai luoghi comuni." stranieriinitalia.it.

Ginsborg, Paul. 2003. *A History of Contemporary Italy: Society and Politics, 1943–1988.* New York: Palgrave Macmillan.

Ginsborg, Paul. 2001. *Italy and Its Discontents: Family, Civil Society, State, 1980–2001.* London: Allen Lane.

Giorgi, Carlo. 2003. "La vita nel tunnel." *Terre di Mezzo*: 17.

Giusti, Mariangela. 2008. *Immigrati e tempo libero.* Novara: De Agostini.

Glennie, Paul, and Nigel Thrift. 1996. "Consumption, Shopping and Gender." In *Retailing, Consumption, and Capital: Towards the New Retail Geography*, ed. Neil Wrigley and Michelle Lowe, 221–37. Harlow: Longman.

Gnoli, Sofia. 2005. *Un secolo di moda Italiana.* Roma: Meltemi.

Goldstein, Daniel. 2010. *The Spectacular City: Violence and Performance in Urban Bolivia.* Durham: Duke University Press.

Goldstein, Matteo Bolocan, and Bertrando Bonfantini, eds. 2007. *Milano incompiuta: Interpretazioni urbanistiche del mutamento.* Milan: Quaderni del Dipartimento di Architettura e Pianificazione.

Goldstein, Matteo Bolocan, Bertrando Bonfantini, and Sivia Botti. 2007. "Ricercare Milano: Tra mercato urbano e sfera pubblica." In *Milano incompiuta: Interpretazioni urbanistiche del mutamento,* ed. Matteo Bolocan Goldstein and Bertrando Bonfantini, 9–16. Milan: Quaderni del Dipartimento di Architettura e Pianificazione.

Gordon, Avery. 1997. *Ghostly Matters: Haunting and the Sociological Imagination.* Minneapolis: University of Minnesota Press.

Granata, Elena, Christian Novak, and Emanuele Polizzi. 2003. "Immigrazione dall' Asia e trasformazione urbana." In *Asia a Milano: Famiglie, ambienti e lavori delle popolazioni asiatiche a Milano,* ed. Daniele Cologna, 96–152. Milan: Abitare Segesta.

Grasseni, Cristina. 2011. "Skilled Visions: Toward an Ecology of Visual Inscriptions." In *Made to Be Seen: Perspectives on the History of Visual Anthropology,* ed. Marcus Banks and Jay Ruby, 19–43. Chicago: University of Chicago Press.

Gregson, Nicky, and Gillian Rose. 2000. "Taking Butler Elsewhere: Performativities, Spatialities and Subjectivities." *Environment and Planning D* 18 (4): 433–52. http://dx.doi.org/10.1068/d232.

Grispigni, Marco. 1994. "Sulla soglia del nomadismo." In *Comunità virtuali, i centri sociali in Italia,* ed. Francesco Adinolfi et al., 23–30. Rome: ManifestoLibri.

Gualmini, Elisabetta, and Maurizio Ferrera. 2004. *Rescued by Europe? Social and Labour Market Reforms in Italy from Maastricht to Berlusconi.* Amsterdam: Amsterdam University Press.

Guano, Emanuela. 2010. "Taxpayers, Thieves, and the State: Fiscal Citizenship in Contemporary Italy." *Ethnos* 75 (4): 471–95. http://dx.doi.org/10.1080/00141844.2010.522246.

Guano, Emanuela. 2008. "Unstable Citizenship: Dealing with Bureaucracies in Contemporary Italy." Lecture held at Simon Fraser University, Burnaby, BC, on 28 March.

Guano, Emanuela. 2007. "Respectable Ladies and Uncouth Men: The Performative Politics of Class and Gender in the Public Realm of an Italian City." *Journal of American Folklore* 120 (475): 48–72. http://dx.doi.org/10.1353/jaf.2007.0011.

Guano, Emanuela. 2003. "A Stroll through La Boca: The Politics and Poetics of Spatial Experience in a Buenos Aires Neighborhood." *Space and Culture* 6 (4): 356–76. http://dx.doi.org/10.1177/1206331203257250.

Guano, Emanuela. 2002. "Ruining the President's Spectacle: Theatricality and Telepolitics in the Buenos Aires Public Sphere." *Journal of Visual Culture* 1 (3): 303–23. http://dx.doi.org/10.1177/147041290200100304.

"Guerra di strada tra cinesi e forze dell'ordine." 2007, 13 Apr. *Corriere della Sera*. www.corriere.it/vivimilano

Habermas, Jürgen 1989. *The Structural Transformations of the Public Sphere.* Cambridge, MA: MIT Press.

Habermas, Jürgen, Sara Lennox, and Frank Lennox. 1974. "The Public Sphere: An Encyclopedia Article (1964)." *New German Critique* 3 (3): 49–55. http://dx.doi.org/10.2307/487737.

Hansen, Peo. 2000. "'European Citizenship,' or where Neoliberalism Meets Ethno-Culturalism." *European Societies* 2 (2): 139–65. http://dx.doi.org/10.1080/146166900412046.

Haraway, Donna Jeanne. 1991. *Simians, Cyborgs, and Women: The Reinvention of Nature.* London: Free Association Books.

Harms, Erik. 2010. *Saigon's Edge.* Minneapolis: University of Minnesota Press.

Harms, Erik. 2009. "Vietnam's Civilizing Process and the Retreat from the Street: A Turtle's Eye View from Ho Chi Minh City." *City & Society* 21 (2): 182–206. http://dx.doi.org/10.1111/j.1548-744X.2009.01021.x.

Herzfeld, Michael. 2009. *Evicted from Eternity: The Restructuring of Modern Rome.* Chicago: University of Chicago Press. http://dx.doi.org/10.7208/chicago/9780226329079.001.0001.

Holston, James, ed. 1999. *Cities and Citizenship.* Durham: Duke University Press.

Holston, James, and Arjun Appadurai. 1999. "Cities and Citizenship." In *Cities and Citizenship*, ed. James Holston, 1–16. Durham: Duke University Press.

Indini, Andrea. 2009, 16 June. "Immigrazione, monito della Lega: 'No ai bivacchi nelle aree verdi.'" *Il Giornale*. http://www.ilgiornale.it/news/immigrazione-monito-lega-no-ai-bivacchi-nelle-aree-verdi.html

Ingold, Tim, and Jo L. Vergunst, eds. 2008. *Ways of Walking: Ethnography and Practice on Foot.* Aldershot, UK: Ashgate.

Isin, Engin, and Kim Rygiel. 2007. "Of Other Global Cities: Frontiers, Zones, Camps." In *Cities of the South: Citizenship and Exclusion in the 21st Century*, ed. Barbara Drieskens, Franck Mermier, and Heiko Wimmen, 177–209. London: SAQI.

Iveson, Kurt. 2007. *Publics and the City.* Oxford: Blackwell. http://dx.doi.org/10.1002/9780470761748.

Jackson, Michael. 2005. *Existential Anthropology: Events, Exigencies and Effects.* New York: Berghahn.

Krause, Elizabeth. 2001. "'Empty Cradles' and the Quiet Revolution: Demographic Discourse and Cultural Struggles of Gender, Race, and Class in Italy." *Cultural Anthropology* 16 (4): 576–611. http://dx.doi.org/10.1525/can.2001.16.4.576.

LaboratorioCittàPubblica. 2009. *Città Pubbliche: Linee guida per la riqualificazione urbana*. Milan: Mondadori.

La Repubblica. 2007, 24 Apr. "Modelle Lavavetri." http://www.repubblica.it.

Lees, Loretta. 2008. "Gentrification and Social Mixing: Towards an Inclusive Urban Renaissance?" *Urban Studies (Edinburgh, Scotland)* 45 (12): 2449–70. http://dx.doi.org/10.1177/0042098008097099.

Lefebvre, Henri. 1991. *The Production of Space*. Oxford: Blackwell.

Leogrande, Alessandro, and Grazia Naletto. 2002. *Bada alla bossi-fini! Contenuti, 'cultura' e demagogia della nuova legge sull'immigrazione*. Milan: Altraeconomia, Asgi, ICS, Lo straniero, Lunaria, Terre di mezzo.

Leoncavallo Website. http://www.leoncavallo.org.

Low, Setha. 2000. *On the Plaza: The Politics of Public Space and Culture*. Austin: University of Texas Press.

Low, Setha, and Neil Smith, eds. 2006. *The Politics of Public Space*. New York: Routledge.

Lutz, Catherine, and Jane Collins. 1993. *Reading National Geographic*. Chicago: University of Chicago Press.

Madison, Soyini. 1999. "Performing Theory/Embodied Writing." *Text and Performance Quarterly* 19 (2): 107–24. http://dx.doi.org/10.1080/10462939909366254.

Malkki, Liisa. 2007. "Tradition and Improvisation in Ethnographic Field Research." In *Improvising Theory: Process and Temporality in Ethnographic Fieldwork*, ed. Allaine Cerwonka and Liisa Malkki, 163–86. Chicago: University of Chicago Press.

Mankekar, Purnima. 1999. *Screening Culture, Viewing Politics: An Ethnography of Television, Womanhood, and Nation in Postcolonial India*. Durham: Duke University Press.

Maritano, Laura. 2004. "Immigration, Nationalism and Exclusionary Understandings of Place in Turin." In *Italian Cityscapes: Culture and Urban Change in Contemporary Italy*, ed. Robert Lumley and John Foot, 61–74. Exeter: University of Exeter Press.

Maritano, Laura. 2002. "An Obsession with Cultural Difference: Representations of Immigrants in Turin." In *The Politics of Recognizing Difference: Multiculturalism Italian-Style*, ed. Jeff Pratt and Ralph Grillo, 59–75. Aldershot: Ashgate.

Martinotti, Guido. 2003, 20 Jan. "Space, Technologies, and Populations in the New Metropolis." Lecture for the European on-Line Seminar on Urban Transformation, Poverty, Spatial Segregation, and Social Exclusion. http://www.shakti.uniurb.it/eurex/; accessed Jan. 2003.

Martinotti, Guido. 1993. *Metropoli: La nuova morfologia sociale della città*. Bologna: Il Mulino.

Mauri, Luigi, Daniele Cologna, Elena Granata, and Christian Novak. 2003. "L'Immigrazione asiatica e la cittá: Potenzialitá e snodi critici per le politiche urbane e sociali." In *Asia a Milano: Famiglie, ambienti e lavori delle popolazioni asiatiche a Milano*, ed. Daniele Cologna, 228–32. Milan: Abitare Segesta.

McLagan, Meg. 2002. "Spectacles of Difference: Cultural Activism and the Mass Mediation of Tibet." In *Media Worlds: Anthropology on New Terrain*, ed. Faye D. Ginsburg, Lila Abu-Lughod, and Brian Larkin, 90–111. Berkeley: University of California Press.

Melossi, Dario. 2003. "'In a Peaceful Life': Migration and the Crime of Modernity in Europe/Italy." *Punishment and Society* 5 (4): 371–97. http://dx.doi.org/10.1177/14624745030054001.

Membretti, Andrea. 2007. "Centro Sociale Leoncavallo: Building Citizenship as an Innovative Service." *European Urban and Regional Studies* 14 (3): 252–63. http://dx.doi.org/10.1177/0969776407077742.

Membretti, Andrea. 2003. *Leoncavallo: Spazio pubblico autogestito*. Milan: Mamme Del Leoncavallo.

Merlino, Massimo. 2009. *The Italian (In)Security Package: Security vs. Rule of Law and Fundamental Rights in the EU*. Changing Landscape of European Liberty and Security papers, research paper no. 14. http://www.ceps.eu.

Merlo, Flavio. 2001. "Un morbido filo di sole: Il cashmere per uomo di Baroni." In *Al di lá della moda: Oggetti, storie, significati*, ed. Lucia Ruggerone, 55–94. Milan: FrancoAngeli.

Merrill, Heather. 2011. "Migrations and Surplus Populations: Race and Deindustrialization in Northern Italy." *Antipode* 43 (5): 1542–72. http://dx.doi.org/10.1111/j.1467-8330.2011.00904.x.

Merrill, Heather. 2006. *An Alliance of Women: Immigration and the Politics of Race*. Minneapolis: University of Minnesota Press.

Mitchell, Don. 2003. *The Right to the City: Social Justice and the Fight for Public Space*. New York: Guilford Press.

Mitchell, Don. 1996. Introduction: Public Space and the City. *Urban Geography* 17 (2): 127–31.

Mitchell, Don. 1995. "The End of Public Space? People's Park, Definitions of the Public, and Democracy." *Annals of the Association of American Geographers* 85 (1):108–33.

Mitchell, Don, and Lynn A. Staeheli. 2005. "Turning Social Relations into Space: Property, Law and the Plaza of Santa Fe, New Mexico." *Landscape Research* 30 (3): 361–78.

Miyazaki, Hirokazu. 2004. *The Method of Hope: Anthropology, Philosophy, and Fijian Knowledge*. Palo Alto: Stanford University Press.

Monteleone, Raffaele, and Lidia Manzo. 2010. "Canonica-sarpi: Un quartiere storico in fuga dal presente." In *Milano Downtown: Azione pubblica e luoghi dell'abitare*, ed. Massimo Brococoli and Paola Savoldi, 133–61. Milan: Et al. Edizioni.

Molé, Noelle. 2012. *Labor Disorders in Neoliberal Italy: Mobbing, Well-Being, and the Workplace*. Bloomington: Indiana University Press.

Molé, Noelle. 2010. "Precarious Subjects: Anticipating Neoliberalism in Northern Italy's Workplace." *American Anthropologist* 112 (1): 38–53. http://dx.doi.org/10.1111/j.1548-1433.2009.01195.x.

"Montenapo, il comune accusa gli stilisti." 2009, 19 Feb. *Corriere della Sera, Corriere Milano*, 1.

Moretti, Cristina. 2011. "The Wandering Ethnographer: Researching and Representing the City through Everyday Encounters." *Anthropologica*, Special Theme Issue: New Directions in Experimental and Engaged Ethnography 53 (2): 245–55.

Moretti, Cristina. 2008a. "A Walk of Two Women: Vision, Gender, and Belonging in Milan, Italy." In *Gender in an Urban World*, ed. Judith DeSena, 53–75. Bingley: Emerald. http://dx.doi.org/10.1016/S1047-0042(07)00003-7.

Moretti, Cristina. 2008b. "Places and Stages: Performing and Narrating the City in Milan, Italy." *Liminalities: A Journal of Performance Studies*, Special Issue on the City 4 (1): 1–46.

Moroni, Primo. 1996. "Un certo uso sociale dello spazio urbano." In *Centri sociali: Geografie del desiderio. Dati, statisstiche, progetti, mappe, divenire – Consorzio Aaster, Centro Sociale Cox 18*, ed. Centro Sociale Leoncavallo and Primo Moroni, 161–87. Milan: Shake Edizioni Underground.

Mudu, Pierpaolo. 2004. "Resisting and Challenging Neoliberalism: The Development of Italian Social Centers." *Antipode* 36 (5): 917–41. http://dx.doi.org/10.1111/j.1467-8330.2004.00461.x.

Muehlebach, Andrea. 2012. *The Moral Neoliberal: Welfare and Citizenship in Italy*. Chicago: University of Chicago Press. http://dx.doi.org/10.7208/chicago/9780226545417.001.0001.

Multiplicity.lab. 2007. *Milano: Cronache dell'abitare*. Milan: Mondadori.

Murer, Bruno. 2003. "Dall' immigrazione virtuale all'immigrazione reale." In *Asia a Milano: Famiglie, ambienti e lavori delle popolazioni asiatiche a Milano*, ed. Daniele Cologna, 10–14. Milan: Abitare Segesta.

Navaro-Yashin, Yael. 2009. "Affective Spaces, Melancholic Objects: Ruination and the Production of Anthropological Knowledge." *Journal of the Royal Anthropological Institute* 15 (1): 1–18. http://dx.doi.org/10.1111/j.1467-9655.2008.01527.x.

Navaro-Yashin, Yael. 2002. *Faces of the State: Secularism and Public Life in Turkey*. Princeton: Princeton University Press.

Narayan, Kirin. 1993. "How Native Is a 'Native' Anthropologist?" *American Anthropologist* 95 (3): 671–86. http://dx.doi.org/10.1525/aa.1993.95.3.02a00070.

Newcomb, Rachel. 2006. "Gendering the City, Gendering the Nation: Contesting Urban Space in Fes, Morocco." *City & Society* 18 (2): 288–311. http://dx.doi.org/10.1525/city.2006.18.2.288.

Naga. 2009. *Razzismi quotidiani*. Milan.

Naga. 2005. *Abitare la città invisibile*. Rapporto 2003 –2004 marzo 2005. Osservatorio Naga, Gruppo Medicina di Strada.

Naga. 2003. *La città invisibile*. Rapporto sulla popolazione delle baraccopoli e delle aree dismesse Milanesi. Milan: Gruppo di Medicina di Strada del Naga.

Nicolini, Kim. 1998. "The Streets of San Francisco: A Personal Geography." In *Bad Subjects: Political Education for Everyday Life*, ed. 20 Bad Subjects Production Team. New York: New York University Press.

Novak, Christian, and Viviana Andriola. 2008. "Milano, lungo Via Padova: Periferie in sequenza." In *Tracce di quartieri: Il legame sociale nella città che cambia*, ed. Marco Cremaschi, 222–48. Milan: FrancoAngeli.

Paltrinieri, Anna Casella. 2001. "Collaboratrici domestiche straniere in Italia: L'Interazione culturale possibile." *Studi emigrazione/Migration Studies* 38 (143): 515–38.

Paone, Sonia. 2008. *Città in frantumi: Sicurezza, emergenza e produzione dello spazio*. Milan: FrancoAngeli.

Parreñas, Rhacel Salazar. 2001. *Servants of Globalization: Women, Migration and Domestic Work*. Stanford: Stanford University Press.

Parsi, Vittorio, and Enrico Tacchi. 2003. *Quarto Oggiaro, Bovisa, Dergano: Prospettive di riqualificazione della periferia di Milano*. Milan: FrancoAngeli.

Partridge, Damani. 2008. "We Were Dancing in the Club, Not on the Berlin Wall: Black Bodies, Street Bureaucrats, and Exclusionary Incorporation into the New Europe." *Cultural Anthropology* 23 (4): 660–87. http://dx.doi.org/10.1111/j.1548-1360.2008.00022.x.

Pasquino, Gianfranco. 2007. "Italian Politics: No Improvement in Sight." *Journal of Modern Italian Studies* 12 (3): 273–85. http://dx.doi.org/10.1080/13545710701455627.

Petrillo, Gianfranco. 2004. "The Two Waves: Milan as a City of Immigration, 1955–1995." In *Italian Cityscapes: Culture and Urban Change in Contemporary Italy*, ed. Robert Lumley and John Foot, 31–45. Exeter: University of Exeter Press.

Pigg, Stacy Leigh. 1996. "The Credible and the Credulous: The Question of 'Villagers' Beliefs' in Nepal." *Cultural Anthropology* 11 (2): 160–201. http://dx.doi.org/10.1525/can.1996.11.2.02a00020.

Pink, Sarah. 2008. "An Urban Tour: The Sensory Sociality of Ethnographic Place-Making ." *Ethnography* 9 (2): 175–96. http://dx.doi.org/10.1177/1466138108089467.

Pink, Sarah. 2007. *Doing Visual Ethnography: Images, Media, and Representation in Research*. London: Sage.

Pinney, Christopher. 2004. *'Photos of the Gods': The Printed Image and Political Struggle in India*. London: Reaktion Books.

Pinney, Christopher. 2002. "The Indian Work of Art in the Age of Mechanical Reproduction: Or, What Happens when Peasants 'Get Hold' of Images." In *Media Worlds: Anthropology on New Terrain*, ed. Faye D. Ginsburg, Lila Abu-Lughod, and Brian Larkin, 355–69. Berkeley: University of California Press.

Potuoğlu-Cook, Öykü. 2006. "Beyond the Glitter: Belly Dance and Neoliberal Gentrification in Istanbul." *Cultural Anthropology* 21 (4): 633–60. http://dx.doi.org/10.1525/can.2006.21.4.633.

Porta Nuova Website. www.porta-nuova.com.

Portelli, Alessandro. 1990. *The Death of Luigi Trastulli, and Other Stories: Form and Meaning in Oral History*. Albany: State University of New York Press.

Pratt, Minnie Bruce. 1988. "Identity: Skin, Blood, Heart." In *Yours in Struggle: Three Feminist Perspectives on Anti-Semitism and Racism*, ed. Elly Bulkin, Minnie Bruce Pratt, and Barbara Smith, 11–63. Ithaca, NY: Firebrand Books.

Pratt, Jeff, and Ralph Grillo, eds. 2002. *The Politics of Recognizing Difference: Multiculturalism Italian-Style*. Aldershot: Ashgate.

Preston, Valerie, and Ebru Ustundag. 2005. "Feminist Geographies of the 'City': Multiple Voices, Multiple Meanings." In *A Companion to Feminist Geography*, ed. Lise Nelson and Joni Seager, 211–27. Oxford: Blackwell.

"Prodi e Fassino Disertano la Fiaccolata." 2006, 17 Mar. *Corriere della Sera*. www.corriere.it.

Quagliata, Livio. 1994. "Ghetto in libera uscita." In *Comunità virtuali, i centri sociali in Italia*, ed. Francesco Adinolfi, et al., 75–81. Rome: ManifestoLibri.

Quassoli, Fabio. 2004. "Making the Neighbourhood Safer: Social Alarm, Police Practices and Immigrant Exclusion in Italy." *Journal of Ethnic and Migration Studies* 30 (6): 1163–81. http://dx.doi.org/10.1080/1369183042000286296.

Rademacher, A. 2008. "Fluid City, Solid State: Urban Environmental Territory in a State of Emergency, Kathmandu." *City & Society* 20 (1): 105–29. http://dx.doi.org/10.1111/j.1548-744X.2008.00008.x.

Razack, Sherene. 2000. "Gendered Racial Violence and Spatialized Justice: The Murder of Pamela George." *Canadian Journal of Law and Society* 15 (2): 91–130.

"Ripamonti, Via Coari: Una nostra lettrice solleva il problema immigrati nei parchi." 2008, 13 May. http://milano.blogosfere.it/post/113286/ripamonti-via-coari-una-nostra-lettrice-solleva-il-problema-immigrati-nei-parchi.

Richardson, Tanya. 2008. *Kaleidoscopic Odessa: History and Place in Contemporary Ukraine.* Toronto: University of Toronto Press.

Rose, Gillian. 1993. *Feminism and Geography: The Limits of Geographical Knowledge.* Minneapolis: University of Minnesota Press.

Roseman, Sharon. 1996. "'How We Built the Road': The Politics of Memory in Rural Galicia." *American Ethnologist* 23 (4): 836–60. http://dx.doi.org/10.1525/ae.1996.23.4.02a00090.

Rotenberg, Robert. 2001. "Metropolitanism and the Transformation of Urban Space in Nineteenth-Century Colonial Metropoles." *American Anthropologist* 103 (1): 7–15. http://dx.doi.org/10.1525/aa.2001.103.1.7.

Rudnyckyj, Daromir. 2011. "Circulating Tears and Managing Hearts: Governing through Affect in an Indonesian Steel Factory." *Anthropological Theory* 11 (1): 63–87. http://dx.doi.org/10.1177/1463499610395444.

Ruggerone, Lucia. 2004. "Il corpo simulato: Immagini femminili nella fotografia di moda." *Studi di Sociologia* 3: 277–305.

Saint-Blancat, Chantal, and Ottavia Schmidt di Friedberg. 2005. "Why Are Mosques a Problem? Local Politics and Fear of Islam in Northern Italy." *Journal of Ethnic and Migration Studies* 31 (6): 1083–1104. http://dx.doi.org/10.1080/13691830500282881.

Saitta, Pietro. 2011 "Neoliberismo e controllo dell'immigrazione: Il fallimento della 'tolleranza zero' e i paradossali esiti dell'informalità." In *Narrare l'altro: Pratiche discorsive sull'immigrazione,* ed. Domenico Carzo, 107–26. Rome: A-racne.

Salazar, Noel. 2011. "The Power of Imagination in Transnational Mobilities: Identities." *Global Studies in Culture and Power* 18 (6): 576–98.

Savoldi, Paola. 2010. "Santa Giulia: Da città d'avanguardia a quartiere periurbano." In *Milano Downtown: Azione pubblica e luoghi dell'abitare,* ed. Massimo Bricocoli and Paola Savoldi, 45–73. Milan: Et al. Edizioni.

Schiavi, Giangiacomo. 2004, 18 Dec. "I trenta-quarantenni, Milano e la voglia di fuga." *Corriere della Sera*: 55.

Schielke, Samuli. 2013. "Engaging the World on the Alexandria Waterfront." In *The Global Horizon: Expectations of Migration in Africa and the Middle East,* ed. Knut Graw and Samuli Schielke, 175–92. Leuven: Leuven University Press.

Schielke, Samuli. 2012. "Surfaces of Longing: Cosmopolitan Aspiration and Frustration in Egypt." *City & Society* 24 (1): 29–37. http://dx.doi.org/ 10.1111/j.1548-744X.2012.01066.x.

Schielke, Samuli. 2008. "Policing Ambiguity: Muslim Saints-Day Festivals and the Moral Geography of Public Space in Egypt." *American Ethnologist* 35 (4): 539–52. http://dx.doi.org/10.1111/j.1548-1425.2008.00097.x.

Segre Reinach, Simona. 2006. "Milan: The City of Prêt-á-Porter in a World of Fast Fashion." In *Fashion's World Cities*, ed. Christopher Breward and David Gilbert, 123–34. Oxford: Berg.

Segre Reinach, Simona. 2005. *La moda: Un'introduzione*. Bari: Laterza.

Segre Reinach, Simona. 1999. *Mode in Italy: Una lettura antropologica*. Milan: Guerini.

Sigona, Nando. 2011. "The Governance of Romani People in Italy: Discourse, Policy and Practice." *Journal of Modern Italian Studies* 16 (5): 590–606. http:// dx.doi.org/10.1080/1354571X.2011.622468.

Smart, Alan, and Josephine. 2003. "Urbanization and the Global Perspective." *Annual Review of Anthropology* 32: 263–85.

Soley-Beltran, Patrícia. 2004. "Modelling Femininity." *European Journal of Women's Studies* 11 (3): 309–26. http://dx.doi.org/10.1177/1350506804044465.

Stacul, Jaro. 2007. "Understanding Neoliberalism: Reflections on the 'End of Politics' in Northern Italy." *Journal of Modern Italian Studies* 12 (4): 450–9. http://dx.doi.org/10.1080/13545710701640814.

Steedly, Mary Margaret. 1993. *Hanging without a Rope: Narrative Experience in Colonial and Postcolonial Karoland*. Princeton: Princeton University Press.

Stewart, Kathleen. 2008. "Weak Theory in an Unfinished World." *Journal of Folklore Research* 45 (1): 71–82. http://dx.doi.org/10.2979/JFR.2008 .45.1.71.

Stewart, Kathleen. 2005. "Cultural Poesis: The Generativity of Emergent Things." In *The Sage Handbook of Qualitative Research*, ed. Norman Denzin and Yvonna Lincoln, 1027–42. Thousand Oaks: Sage.

Stewart, Kathleen. 1996. *A Space on the Side of the Road: Cultural Poetics in an 'Other' America*. Princeton: Princeton University Press.

Taylor, Diana. 1997. *Disappearing Acts: Spectacles of Gender and Nationalism in Argentina's 'Dirty War.'* Durham, London: Duke University Press.

TempoRiuso. 2009. *Information Pamphlet*. Milan: Author.

Terre di Mezzo Website. http://www.terre.it.

Terre di Mezzo. 2003. "100 Mesi di Libera Informazione." *Terre di Mezzo*: 100.

Teelucksingh, Cheryl, ed. 2006. *Claiming Space: Racialization in Canadian Cities*. Waterloo: Wilfrid Laurier University Press.

Thrift, Nigel. 2003. "Performance and ... " *Environment & Planning A* 35 (11): 2019–24. http://dx.doi.org/10.1068/a3543a.

Tonnelat, Stephane. 2008. "Out of Frame: The Invisible Life of Urban Interstices – a Case Study in Carenton-le-Pont, Paris, France." *Ethnography* 9 (3): 291–324. http://dx.doi.org/10.1177/1466138108094973.

Tsing, Anna Lowenhaupt. 2005. *Friction: An Ethnography of Global Connections*. Princeton: Princeton University Press.

Tsing, Anna Lowenhaupt. 2000. "The Global Situation." *Cultural Anthropology* 15 (3): 327–60. http://dx.doi.org/10.1525/can.2000.15.3.327.

Tsing, Anna Lowenhaupt. 1993. *In the Realm of the Diamond Queen*. Princeton: Princeton University Press.

Tupeshka, Tammie. 2001. "The Power of Mapping and the Politics of Place: Community Mapping in Vancouver's Grandview Woodland." M.A. thesis, Simon Fraser University.

Urban Centre. 2009. *Schede delle aree in trasformazione*. Municipality of Milan.

Urban Centre. 2008. *La città che sale*. Milan: Broad-way.

Van Aken, Mauro, ed. 2008. *Rifugio Milano: Vie di fuga e vita quotidiana dei richiedenti asilo*. Rome: Carta.

Van Watson, William. 1989. *Pier Paolo Pasolini and the Theatre of the Word*, vol. 60. Ann Arbor: UMI Research Press.

Vanni, Ilaria, and Marcello Tarì. 2005. "On the Life and Deeds of San Precario, Patron Saint of Precarious Workers and Lives." *Fiberculture Journal*, 5: no page numbers.

Vecchi, Benedetto. 1994. "Frammenti di una diversa sfera pubblica." In *Comunità virtuali, i centri sociali in Italia*, ed. Francesco Adinolfi et al., 5–14. Rome: ManifestoLibri.

Verga, Rossella. 2005, May. "Basta feste e degrado, bisogna difendere i parchi." *Corriere della Sera*: 51.

VivereMilano. 2005. Unpublished Manifesto.

Waquant, Loïc. 2001. "The Penalization of Poverty and the Rise of Neoliberalism." *European Journal on Criminal Policy and Research* 9 (4): 401–12. http://dx.doi.org/10.1023/A:1013147404519.

Weszkalnys, Gisa. 2010. *Berlin, Alexanderplatz: Transforming Place in a Unified Germany*. New York: Berghahn.

Wilson, Elizabeth. 1987. *Adorned in Dreams: Fashion and Modernity*. Berkeley: University of California Press.

Wolbert, Barbara. 2001. "The Visual Production of Locality: Turkish Family Pictures, Migration and the Creation of Virtual Neighbourhood." *Visual Anthropology Review* 17 (1): 21–35. http://dx.doi.org/10.1525/var.2001.17.1.21.

Yanagisako, Sylvia. 2002. *Producing Culture and Capital: Family Firms in Italy*. Princeton: Princeton University Press.

Zajczyk, Francesca. 2005. "Segregazione spaziale e condizione abitativa." In *La povertà come condizione e come percezione: Una survey a Milano*, ed. David Benassi, 53–88. Milan: FrancoAngeli.

Zajczyk, Francesca, and Guido Cavalca. 2006. "Povertá al femminile in un contesto ricco: La donna a Milano tra instabilitá occupazionale, abitativa e familiare." In *L'Urbanitá delle donne, creare, faticare, governare*, ed Antonietta Mazzette, 105–28. Milan: FrancoAngeli.

Zontini, Elisabetta. 2004. "Immigrant Women in Barcelona: Coping with the Consequences of Transnational Lives." *Journal of Ethnic and Migration Studies* 30 (6): 1113–44. http://dx.doi.org/10.1080/1369183042000286278.

Zukin, Sharon. 1995. *The Cultures of Cities*. Cambridge: Blackwell.

Index

Plastic Houses (Case di Plastica), 124
Porta Venezia, 19, 288
posters, 37–8, 48–9, 102, 121, 246–8
precariousness, and precarious
labour, 67, 85, 88, 91–2, 124–6,
135, 182
public space(s): and hope, 104–5;
and immigrants, 133, 135–49,
157–8, 166, 184, 188–90; and
performance, 11–14, 81, 96,
139, 201–5, 226–30, 233–4; and
politics, 85–7, 95–6, 102–5,
117–21, 130, 229–30; and stores,
40–5; as constraining, 137–8; as
ethnographic possibility, 32–3;
as field of vision, 8–10, 34–59,
159–72, 199, 237–8; as gathering
place, 139–41, 143, 145; as site of
judgment, 142–6, 217; conflicts in,
4–6, 227–9; construction of, 153–8,
236; exclusion and inequality in,
30–1, 92–5, 97–8, 240; occupied,
107–8, 116–17, 121, 123, 130, 180;
police raids in, 138–9; positive
aspects of, 3–4, 28–9, 89, 147,
225–7; shifting definitions of, 7–8,
27, 31–3, 236–7. See also Duomo
Piazza; parks; piazzas; rallies;
various piazzas by name

race, 5, 9, 24, 30, 57, 89, 120, 133, 136,
142–3, 184, 192, 195–6. See also
visible minority
rallies, 4, 10, 37, 46, 227–9, 231–4
referendum, 100, 103
research "at home," 18
Riva, 8–10, 17, 157
Roma and Sinti people, 50, 59, 74–5,
107, 165, 167–8
Rotenberg, Robert, 52–3

San Precario, 69, 125–6, 129–31,
180–2. See also Serpica Naro
sand dune, public space as, 136, 154,
158, 236
Salvadorians, and Salvadorian-
Italians, 137, 139–40, 144
Scala: Piazza, 19, 63, 65, 184, 166;
Theatre, La, 83, 92, 119, 194, 221
Schuster Centre, 19–20, 137, 139–40
Schuster Youth, 15, 17, 136–40,
142–8, 150, 166
security, 30–1, 37, 73–5, 99, 106,
167, 240
Security Package (*pacchetto
sicurezza*), 37–8, 167, 252
Segre Reinach, Simona, 168–71
Senegalese, and Senegalese-Italians,
201–2, 207–8, 210, 219
Serpica Naro, 181–2, 253. See also San
Precario
Sforza Castle, 19, 138, 188, 209,
219–20, 223
signs and placards, 39, 93–4, 231–4
skyline, 53, 241–2
Social Centres, 14–18, 106–32, 163,
180, 227–8, 231. See also Casa
Loca; Leoncavallo; Stecca degli
Artigiani
southern Italians, 12, 62
squatting, 112, 114–17, 123–4, 130
squeegees, 58–9
Stecca degli Artigiani, 16, 19, 131–2,
175–80, 242–3, 246–8
storyteller in the Duomo Piazza, 29
stranger, 13, 17, 37, 136, 148, 154–5,
158, 227
street vendors, 30, 65, 138, 166–7,
201–2, 213, 221–5
struscio, 54–7, 169–70, 199, 237, 252,
257n6

ANTHROPOLOGICAL HORIZONS

Editor: Michael Lambek, University of Toronto

Published to date:

An Irish Working Class: Explorations in Political Economy and Hegemony, 1800–1950 / Marilyn Silverman (2001)

The Double Twist: From Ethnography to Morphodynamics / Edited by Pierre Maranda (2001)

The House of Difference: Cultural Politics and National Identity in Canada / Eva Mackey (2002)

Writing and Colonialism in Northern Ghana: The Encounter between the LoDagaa and the 'World on Paper,' 1892–1991 / Sean Hawkins (2002)

Guardians of the Transcendent: An Ethnography of a Jain Ascetic Community / Anne Vallely (2002)

The Hot and the Cold: Ills of Humans and Maize in Native Mexico / Jacques M. Chevalier and Andrés Sánchez Bain (2003)

Figured Worlds: Ontological Obstacles in Intercultural Relations / Edited by John Clammer, Sylvie Poirier, and Eric Schwimmer (2004)

Revenge of the Windigo: The Construction of the Mind and Mental Health of North American Aboriginal Peoples / James B. Waldram (2004)

The Cultural Politics of Markets: Economic Liberalization and Social Change in Nepal / Katherine Neilson Rankin (2004)

A World of Relationships: Itineraries, Dreams, and Events in the Australian Western Desert / Sylvie Poirier (2005)

The Politics of the Past in an Argentine Working-Class Neighbourhood / Lindsay DuBois (2005)

Youth and Identity Politics in South Africa, 1990–1994 / Sibusisiwe Nombuso Dlamini (2005)

Maps of Experience: The Anchoring of Land to Story in Secwepemc Discourse / Andie Diane Palmer (2005)

Beyond Bodies: Rain-Making and Sense-Making in Tanzania / Todd Sanders (2008)

We Are Now a Nation: Croats between 'Home' and 'Homeland' / Daphne N. Winland (2008)

Kaleidoscopic Odessa: History and Place in Post-Soviet Ukraine / Tanya Richardson (2008)

Invaders as Ancestors: On the Intercultural Making and Unmaking of Spanish Colonialism in the Andes / Peter Gose (2008)

From Equality to Inequality: Social Change among Newly Sedentary Lanoh Hunter-Gatherer Traders of Peninsular Malaysia / Csilla Dallos (2011)

Rural Nostalgias and Transnational Dreams: Identity and Modernity among Jat Sikhs / Nicola Mooney (2011)

Dimensions of Development: History, Community, and Change in Allpachico, Peru / Susan Vincent (2012)

People of Substance: An Ethnography of Morality in the Colombian Amazon / Carlos David Londoño Sulkin (2012)